Pathway to Regeneration

An Astrophysicist's Journey into Food, Health, Climate, and Complexity

Phil Gregory

**Professor Emeritus
Physics and Astronomy
University of British Columbia
Vancouver, Canada**

Copyright © 2021 Philip C. Gregory

All rights reserved, including the right to reproduce this book or portions thereof in any form whatsoever, without the written permission of the author.

This book contains the opinions and ideas of the author. It is intended to provide helpful and informative material on the subjects addressed in the book. It is distributed with the understanding that the author is not engaged in rendering medical, health, or any other kind of personal professional services in this book. The reader should consult his or her medical, health, or other competent professional before adopting any of the suggestions in this book or drawing inferences from it.

Mention of any specific companies, organizations, or authorities in this book does not imply endorsement by the author, nor does mention of specific companies, organizations, or authorities imply that they endorse this book or its author.

Internet addresses given in this book were accurate at the time of writing. The author is not responsible for the content of websites he does not own.

The author specifically disclaims all responsibility for any liability, loss, or risk, personal or otherwise, that is incurred as a consequence, directly or indirectly, of the use and appreciation of any of the contents of this book.

All trademarks within this book belong to their legitimate owners.

Published by:
Philip Christopher Gregory
1380 Adams Rd
Bowen Island, BC, V0N 1G2
Canada

This book was created using free software: LibreOffice Writer 6.2.8.2 (word processing, page layout and PDF production) and GIMP 2.10.22 (image editing).
First published March 2021
Reprinted June 2021

Cover photograph by NASA: Earthrise captured by NASA's Lunar Reconnaissance Orbiter

ISBN 9781777500634

Subjects: LCSH: Sustainable agriculture. | LCSH: Food security. | LCSH: Soil microbial ecology. | LCSH: Climate change mitigation. | LCSH: Agricultural chemicals - Health aspects.

Table of Contents

Introduction...1
PART 1 CONTEXT..2
   1. Starting point, personal health...3
   2. Turning point, agriculture's looming collapse...5
   3. The green revolution..6
PART 2 THE SOIL BIOLOGY REVOLUTION AND NATURE'S COMPLEXITY.....9
   4. We have been converting living soil to dirt..10
   5. Nature's barter system..12
   6. Our choice: put more carbon into the atmosphere or sequester it in the soil.........15
   7. The soil carbon sponge..17
   8. Saskatchewan no-till farmers...19
   9. What are plants made of?...21
   10. Exporting soil nutrients..22
   11. Nutritional declines in foods..26
   12. The browning of the green revolution...29
   13. Nature's complexity is amazing...31
   14. The new scoop on methane..35
   15. Human planning versus nature's complexity...37
PART 3 REGENERATIVE AGRICULTURE, A SUSTAINABLE FUTURE...............39
   16. Regenerative agriculture, sustainability is not enough............................40
   17. Six principles of regenerative agriculture..42
   18. The global thermostat and how to cool the planet quickly.....................44
   19. The ABCD of agriculture...48
   20. One man's journey through agriculture...50
   21. The true cost of food..55
   22. Bountiful small regenerative farms..58

PART 4 IMPACT OF FOOD PRODUCTION ON HUMAN HEALTH 62
23. The hidden microbial world 63
Soil food web and soil microbiome 63
Human microbes and the human microbiome 63
24. Effect of antibiotics on our microbiome 64
25. Glyphosate sprayed on crops just before harvesting 66
Glyphosate, leaky gut, chronic disease, and a possible new medical paradigm 66
26. How does glyphosate kill 69
27. The rise of chronic diseases 71
28. Processed food diseases 74
29. Glyphosate in breast milk and pesticide safety issue 77
30. Fake science on trial in the courts 80
31. Whose interests do health regulators serve? 82
PART 5 SUMMARY AND CONCLUSIONS 84
32. Summary 85
33. So what can we do? 88
34. How do I feel, knowing what I now know? 92
References 94
Acknowledgements 108
Alphabetical Index 109

Introduction

I dream of a day when I can joyfully accept a dinner invitation for a meal you lovingly put together, without having to inquire about the source of your food.

How did I arrive at this point where I see much of our food today as toxic to animal, human, and environmental health. But wait a minute, shouldn't everyone be concerned about the source of their food? I imagine you have heard that you are what you eat. It turns out that science is now providing a sound basis for this belief, perhaps just in time, because we have been spraying more and more toxic chemicals under the slogan of feeding the world. Under this industrial agriculture paradigm, referred to as the 'Green Revolution', many farmers wake up asking what chemical they are going to use today and how can they keep their children safe from the spray.

Now you are beginning to get a taste of my life or more correctly our lives, as my wife, Jackie, is on the same page. We have been on a journey of health discovery for 40 years all the while maintaining active careers in teaching and research. Although the starting point was personal health, the journey led me into agriculture, soil biology, climate change, and complexity.

Just as you start to think you have the measure of this book get ready to be surprised. The book continues to build to conclusions that might appear fantastical to many if they were simply stated at the beginning. Simply put, life as normal is leading us in the direction of extinction but it doesn't have to be that way. We need a compass to show us the way to transform human activity into a cooperating 'Team Nature' player. Finding that compass is what this book is all about and as expected it involves a major paradigm shift that is arrived at through an exploration of the interconnections between agricultural practices, soil, plants, microbes, animals (including us), and the planet.

I have tried to keep this account as short as possible, aimed at a wide audience including family, friends and inquiring individuals from all walks of life. I provide lots of references including articles, interviews, YouTube videos, reports, and peer reviewed scientific publications. I have ventured into many new silos of knowledge far from my own, benefiting from my 50 years of research experience and with ready access to the world's literature through my university library. Along the way, I have kept my eyes open for interesting outliers to mainstream thinking, because as Einstein famously said "We can't solve problems by using the same kind of thinking we used when we created them."

PART 1
CONTEXT

1. Starting point, personal health

I started this journey late in my 30s. I was a young professor at the University of British Columbia (UBC) in the Physics and Astronomy Department. For years I had suffered occasional migraines but they became more frequent and I was experiencing a lot of brain fog and occasional intestinal pain. As an academic, brain fog is incapacitating. Imagine, I would have to read each sentence three times for it to register. I decided to go off coffee and tea because of intestinal pains and their negative effect on my sleep. I started having apple juice instead and took pride in finding the best price. I look back in horror from my current state of knowledge, because apples are regularly high in the so-called "Dirty Dozen," the 12 fruits and vegetables with the highest concentrations of pesticides. To think I was drinking the cheapest variety! I was also unaware of the health concerns over consuming fructose without fibre that are discussed in Chapter 28. Along the way I became aware that all was not well in the food industry from reading *Diet for a New America* by John Robbins. Later, we became fans of Michael Pollan's books like the *Omnivores Dilemma*.

The brain fog I was experiencing made it very difficult to prepare and give lectures. I did the rounds of doctors and specialists and they all declared that they couldn't identify any problem. Meanwhile Jackie was having her own health issues. She decided to try out a naturopathic doctor which I have to admit I was very skeptical about at the time. On her first visit, I received a call to come and collect her because she had just fainted. When I arrived she assured me that she was fine apart from a migraine.

Her visits to the naturopath were clearly of great benefit because she rapidly lost her excess weight and her energy level soared by eating whole foods and more organic. I went knocking on the naturopath's door and came away with lots of interesting reading about yeast infections and testing for food sensitivities. I tried out an elimination diet, where you go off all the foods you regularly eat, which left me with some vegetables and rice. After five days, I started feeling absolutely amazing. My brain was firing on all eight cylinders. I was cured! It was really a black and white change. I was clearly reacting to many of the foods I was eating. Once I went back to my regular diet the fog and migraines came right back. The slow disciplined process of only adding one new food back into my diet every two days took another three years with many minor setbacks. One of my most challenging food sensitivities is black pepper because it is in so many products.

I came across frequent mention of a leaky gut being responsible for food sensitivities and that it was due to an overgrowth of yeast which was making my gut lining porous. I tried repeated yeast kills followed by probiotics and did notice temporary improvements that would last for about six weeks. With leaky gut we become sensitive to the foods we eat most commonly so I started rotating my diet. I don't eat the same thing again for at least three days. Since I adopted this approach, I have not developed any more food sensitivities, but it requires considerable discipline to keep this up. Fortunately, Jackie is very supportive but it often means we are each cooking our own but different meals. Along the way we discovered several very helpful doctors and naturopaths.

Our kids are now accustomed to the fact that when we visit we usually bring our own meal or provide many organic ingredients for the Christmas dinner, for example. Some of them have their own food sensitivity challenges and it is becoming much more common in the population. Restaurants are generally accommodating but it is still the rare restaurant that offers all organic, so we eat out very infrequently.

Over the years I have learned to manage my health by eating all organic and plenty of garlic (a natural yeast killer). I have also noticed significant benefits from using a digestive supplement and probiotics. I am particularly pleased with my mental acuity because I so much enjoy my academic interests. The eating and drinking side of our social life has changed fundamentally. It was too painful to go backwards, so we started training our friends that we would prepare and bring our own meals to any social function we attended. We were only light drinkers before but now we have abandoned alcohol altogether. I had discovered it was one of my migraine triggers.

It was especially challenging for me to attend conferences because a lot of the bonding with other scientists goes on over meals. I always have to find accommodation which enables me to cook. Juggling shopping, food preparation and attending a busy conference in a foreign country is a big challenge. In the long run being a food sensitive canary has increased my consciousness of the growing toxicity of the modern world.

Chapter summary:
- My wife and I learn that our main health issues are food related.

2. Turning point, agriculture's looming collapse

Forty years have gone by and I am now 79. I have been retired now for almost 20 years but I am still active in research and affiliated with UBC. Until six years ago, I was immersed in one of the most exciting adventures in astronomy, namely the discovery of thousands of planets orbiting neighbouring stars in the Milky Way. I was looking forward to the discovery of life on another planet in perhaps the next few decades. However, there was a problem, at least one that was new to my consciousness.

"Only 60 Years of Farming Left If Soil Degradation Continues." This announcement by the UN Food and Agriculture Organization appeared in Scientific American[1] on 5 Dec. 2014 and the world carried on, but not quite. I was sufficiently shocked by the announcement that I decided to set aside my astrophysics research and carry out my own independent investigation to see what, if anything, could be done. I like to think that I transitioned from exoplanet research to exploring what is necessary to sustain human life on planet Earth.

The importance of my decision was reinforced in 2017 when the UK Environment Minister warned[2] that the UK is 30-40 years away from eradication of soil fertility. You have to wonder what our situation will be like once half that time has elapsed.

My decision led me on a fascinating six-year journey into current agricultural practices, soil biology, desertification, animal grazing, climate change, and plant and human health. I learned about some amazing advances that have been made in the last 20 to 30 years and especially in the arena of soil biology, and understanding nature's complexity. I first encountered the soil biology piece on a USDA Natural Resources Conservation Service website[3]. This led me to the work of Dr. Elaine Ingham, one of the pioneers of this soil biology revolution[4]. As part of this journey I completed Dr. Ingham's four foundational courses on the Soil Food Web[5].

Along the way I kept running into the interconnections between agricultural practices, desertification, soil health, human health, and climate change. These connections are a central theme of this book.

Chapter summary:
- There are strong interconnections between agricultural practices, desertification, soil health, human health, and climate change.

3. The green revolution

The backdrop for my journey of discovery is the so-called green revolution that emerged in the 1960s. The term 'Green Revolution' was coined by William Gaud, the Director of the United States Agency for International Development. He was describing the spectacular increases in cereal crop yields that were achieved in developing countries during the 1960s. The key to this revolution were new plant varieties which fully utilised fossil-fuel based fertilisers and other new agrochemicals that had become available during this period.

A good example of this is the work of Norman Ernest Borlaug (1914 – 2009) who is often called "the father of the Green Revolution," and is credited with saving over a billion people worldwide from starvation[6]. He was awarded the Nobel Peace Prize in 1970 in recognition of his contributions to world peace through increasing the food supply.

Norman Borlaug was an American agronomist who led initiatives worldwide that contributed to the extensive increases in agricultural production. Borlaug received his B.S. in forestry in 1937 and PhD in plant pathology and genetics from the University of Minnesota in 1942. He took up an agricultural research position in Mexico, where he instigated a wheat-breeding program that developed semi-dwarf, high-yield, disease-resistant wheat varieties[7]. According to writer and historian, Reay Tannahill, when traditional plants are heavily fertilized with nitrogen they shoot up to an unnatural height and then collapse[8]. Borlaug led the introduction of high-yielding semi-dwarf varieties combined with modern agricultural production techniques to Mexico, Pakistan, and India. When planted using improved irrigation and crop management techniques, these new varieties gave dramatic increases in yield. By 1963, 95% of Mexico's wheat crops used the semi-dwarf varieties developed by Borlaug. According to the UN Food and Agricultural Organization, the wheat harvest in Mexico increased by a factor of nearly six times from 1950-2004 [9]. In India and Pakistan the increases were a factor of four and three, respectively.

Following the success of the Mexican program on wheat, the International Rice Research Institute (IRRI) in the Philippines was created, funded jointly by the Rockefeller and Ford Foundations, which succeeded in producing new varieties of rice with increased yields[6].

In spite of these remarkable advances, multiple problems have come into focus. In the developed countries crop yields led to "food mountains" but at the expense of very high inputs with the consequence of very high environmental impacts. The green revolution has been criticized for bringing large-scale monocultures and fossil-fuel based input-intensive farming techniques to countries that had previously relied on subsistence farming, and for widening social inequality owing to uneven food distribution. The pollution of the atmosphere and water systems caused by this input-intensive farming has become a major global environmental and economic concern[10]. There are also concerns about the long-term sustainability of farming practices encouraged by the green revolution in both the developed and developing world. Figure 1 shows a graph of Canadian gross farm revenue and net farm income from 1929 to 2019 provided to me by Darrin Qualman, author of *Civilization Critical: Energy, Food, Nature, and the Future*[11].

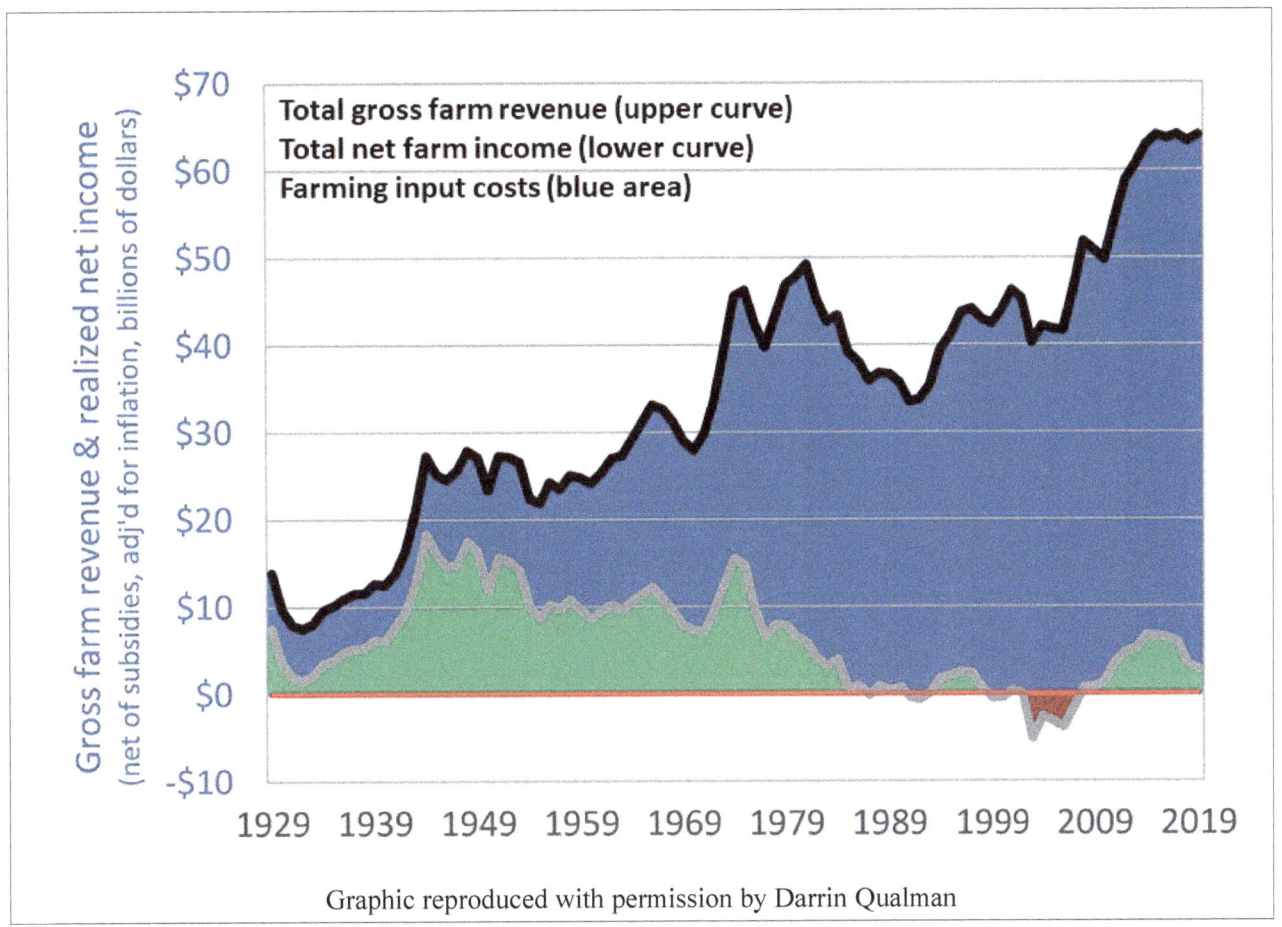

Figure 1. Canadian net farm income and gross revenue, inflation adjusted, net of government payments, 1929–2019. In both the gross farm revenue and net income, taxpayer-funded farm support payments are subtracted, to remove the masking effects these payments can otherwise create. Data is from new Statistics Canada Table numbers, esp. 32-10-0045-01, 32-10-0277-01, 32-10-0052-01, and 32-10-0106-01.

According to Qualman[12], "In the 35-year period from 1985 to 2019, farm-input suppliers of fertilizers, seeds, fuels, pesticides, equipment, banking, accounting services, etc., captured 97% of agricultural revenue, amounting to $1.59 trillion out of $1.63 trillion. These globally dominant transnational corporations have made themselves the primary beneficiaries of the vast food wealth produced on Canadian farms. These companies have extracted almost all the value in the 'value chain'. They have left Canadian taxpayers to backfill farm incomes (approximately $119 billion has been transferred to farmers since 1985). And they have left farmers to borrow the rest (farm debt is now at $115 billion). The massive extraction of wealth by some of the world's most powerful corporations is the cause of an ongoing farm income crisis."

There is also a growing awareness that high synthetic nitrogen fertilizer rates offer only temporary relief leading to a long-term decline in soil productivity (see chapter on 'The Browning of the Green Revolution'). My interpretation of Figure 1 is that the green revolution has pushed farming into a life support situation. According to Qualman, we have been replacing nature's highly diverse circular flows with linear systems where we push huge quantities of fossil-fuel based inputs in at one end and push huge quantities of food out the other end along with greenhouse gases, eroded soils, chemical runoff, toxicity, depletion, loss, and extinction[11].

Now the staff of the Norman Borlaug Institute for Plant Science Research are committed to developing low input, low environmental impact, high yield, and high quality strains of crops. Borlaug believed that to accomplish this goal both conventional technology and biotechnology involving genetic engineering were needed[13]. Is there another approach?

Chapter summary:
- The term "Green Revolution" was coined in the 1960s in response to a spectacular rise in the production of cereal crops in the developing world from the 1950s to the 1990s, through the efforts of plant breeding and synthetic fertilizers and pesticides.
- Norman Borlaug is widely considered to be the father of this revolution through his successful initiative to breed new semi-dwarf, high-yield, disease-resistant wheat varieties. He was awarded the Nobel Peace Prize in 1970 in recognition of his contributions to world peace through increasing the food supply.
- The green revolution has been widely criticized for bringing large-scale monocultures and fossil-fuel based input-intensive farming techniques that are unsustainable and give rise to additional greenhouse gas emissions, eroded soil, chemical runoff, toxicity, depletion, loss, and extinction.
- The green revolution has transferred essentially all the revenue from the farmers to the corporate input providers and put farming into a life support situation.

PART 2
THE SOIL BIOLOGY REVOLUTION AND NATURE'S COMPLEXITY

4. We have been converting living soil to dirt

Here are some of the initial surprises I came across on my journey. The common plow is responsible for turning living soil into dirt. Healthy soil is teeming with billions of microbes in every teaspoon. The microbes include bacteria, fungi, and a host of their microscopic predators. The bacteria and fungi, using carbon based biotic glues and fungal strands (called hyphae), create soil structures which are like underground cities to live and work in. The doors and windows in their soil buildings allow air and water to penetrate to great depths, allowing plant roots to grow up to 20 ft deep in some healthy soils. There is a stunning diversity of these microbes in each gram of healthy soil.

The microbes are nature's way of providing all the additional nutrients that plants require without the use of fossil-fuel based chemicals. Apart from their role in recycling dead plant and animal matter, bacteria and fungi secrete enzymes and organic acids that allow them to extract all the other nutrients plants require from the rocks, sand, silt and clay, as well as nitrogen from the atmosphere. We only recently learned that the world's largest mining operation is run by special types of fungal microbes called mycorrhizae[14]. Mycorrhizal fungi come into direct contact with plant roots and effectively extend the root area for extracting nutrients and water by many hundreds of times. Their microscopic fungal hyphae, literally tunnel through the rocks, sand, silt, and clay and extract all the elements they are made of and provide them to the plants in just the right proportions. We can see their mining tunnels under a microscope[15].

Through the work of researchers like Professor Suzanne Simard of the University of British Columbia, we now know that mycorrhizal fungal networks can link plants together in a Wood Wide Web allowing them to exchange signals as well as nutrients[16][17].

All this changes when we plow or till the soil. This slices and dices the fungal strands and soil structure held together with biotic glues, devastating the homes of the microbes. Without the soil structure, soil is easily washed away or blown away by winds.

What depth of soil are we losing each year? Perhaps the most detailed study of the scientific literature on soil erosion was carried out by the geomorphologist, Professor David Montgomery, of Washington University[18]. He found a global average erosion rate of 3.9 mm/year with an uncertainty of ± 0.3 mm/year. He recommends using the median value of 1.54 mm/year instead of the larger average value. In the case of the median value, 50% of the erosion measurements are less than this value and 50% greater. He also found that the erosion rate drops to 0.08 mm/year for conservation agriculture which minimizes soil disturbance and drops to 0.01 mm/year for native soils.

For every ton of food produced by conventional agriculture we lose approximately 7 tons of topsoil[18][19][20]. In addition, for every ten calories of input energy we put into the food system, we only deliver about one calorie for human nutrition[21][22]. Synthetic nitrogen fertilizer production uses large amounts of natural gas and some coal, and can account for more than 50 per cent of total energy use in conventional agriculture[23]. Since we need to get off of fossil fuels, the industrial agricultural model clearly has no long-term future.

According to Dr. Elaine Ingham, the so-called green revolution, the use of fossil-fuel based fertilizers, simply reflects the damage we have done to our soils, initially by plowing and tilling. Without the microbes to provide the nutrients plants require and build soil structure, soil becomes dirt and then the only way to grow plants is to add chemicals. We just haven't understood nature's important role for the microbes. Fortunately, we can restore the microbes to the soil by inoculating the soil with good compost and by spraying with a compost extract or tea, made from the compost. It is important to ensure the compost is teeming with a good selection of soil microbes. The simplest way of doing that is with a soil microscope.

Chapter summary:
- Healthy soil is teeming with billions of microbes in every teaspoon. The microbes include bacteria, fungi, and a host of microscopic predators. They run extensive mining and recycling operations.
- Mycorrhizal fungal networks can link plants together in a Wood Wide Web allowing them to exchange signals as well as nutrients.
- The bacteria and fungi, using biotic glues and fungal strands, create soil structure with lots of voids to hold air and water.
- Plowing or tillage collapses soil structure built by soil microbes turning living soil into dirt. The damage to soil biology/fertility from frequent plowing helped give rise to the so-called green revolution as a way to grow plants in dirt with agricultural chemicals.
- Current agriculture is not sustainable because without the biotic glues and fungal strands soil is easily eroded away. We are losing seven tons of soil for each ton of food produced.
- We can restore the microbes by inoculating the soil with first class compost and stopping tilling.

5. Nature's barter system

Another important part of the soil biology revolution is the discovery that plants release up to 40% of the sugars they produce from photosynthesis directly through their roots, to attract and feed the soil microbes. Photosynthesis actually captures sunlight energy and stores it in the carbon bonds of sugars. These carbon compounds which are released through the plant roots are called root exudates (see Figure 2). They provide the energy source to run the microbe recycling and mining operations that provide the other elements that plants need to grow strong and healthy. The number of elements considered important to plant growth has steadily grown over time and is now around 20. Some 60 chemical elements are found in the human body, but what all of them are doing there is still unknown[24]. Some 33 elements, including trace and ultratrace elements, are known to be important for biological molecules[25]. Since you are what you eat it is quite possible that the number of elements important to plants will continue to grow as our knowledge continues to advance.

Photosynthesis provides the major elements of carbon, oxygen and hydrogen, and certain bacteria also extract nitrogen from the atmosphere and convert it to plant available ammonium and nitrate. As mentioned above, the microbes are able to mine the other elements from the rocks, sand, silt, and clay. This is nature's barter system in which the plants provide the microbes with sugars in return for all the other elements plants require (see Figure 3).

In this barter system, it is important to acknowledge the crucial role of the microscopic predators of the bacteria and fungi which include protozoa, nematodes, and microarthropods. Most of the bacteria and fungi store the nutrients they extract in high concentrations in their bodies because these are the nutrients they need for their life. Their microscopic predators do not require these nutrients in such high concentrations and so when they eat the bacteria and fungi they poop out the excess in a plant available form. This is all happening right next to the plant roots.

In addition, there is a hierarchy of larger predators including springtails, mites, millipedes, centipedes, fly larvae, beetles, earthworms, spiders, and burrowing animals. In nature high biodiversity translates into population stability and resilience. Unfortunately, in chemical-intensive agriculture, plants soon become addicted to the nitrogen, phosphorus, and potassium fertilizers and stop providing the carbon compounds (root exudates) to power the microbes. The plants are weakened from not receiving the full complement of elements in just the right proportions nature intended. The plants become susceptible to insects and disease, leading to the need for pesticides to fight off nature's garbage collectors. The chemical route produces plants that look like plants but are deficient in many trace elements.

Since the introduction of genetically engineered crops in the 1990s, there has been a massive increase in the use of chemical herbicides that are chelators[27]. These bind to metals like zinc, magnesium, manganese, iron, and copper making them unavailable to the growing plant. The consensus appears to be that you need to eat many oranges today to get the same nutrients that your grandparents got from one orange when they were young[28][29]. See more on this in Chapter 11.

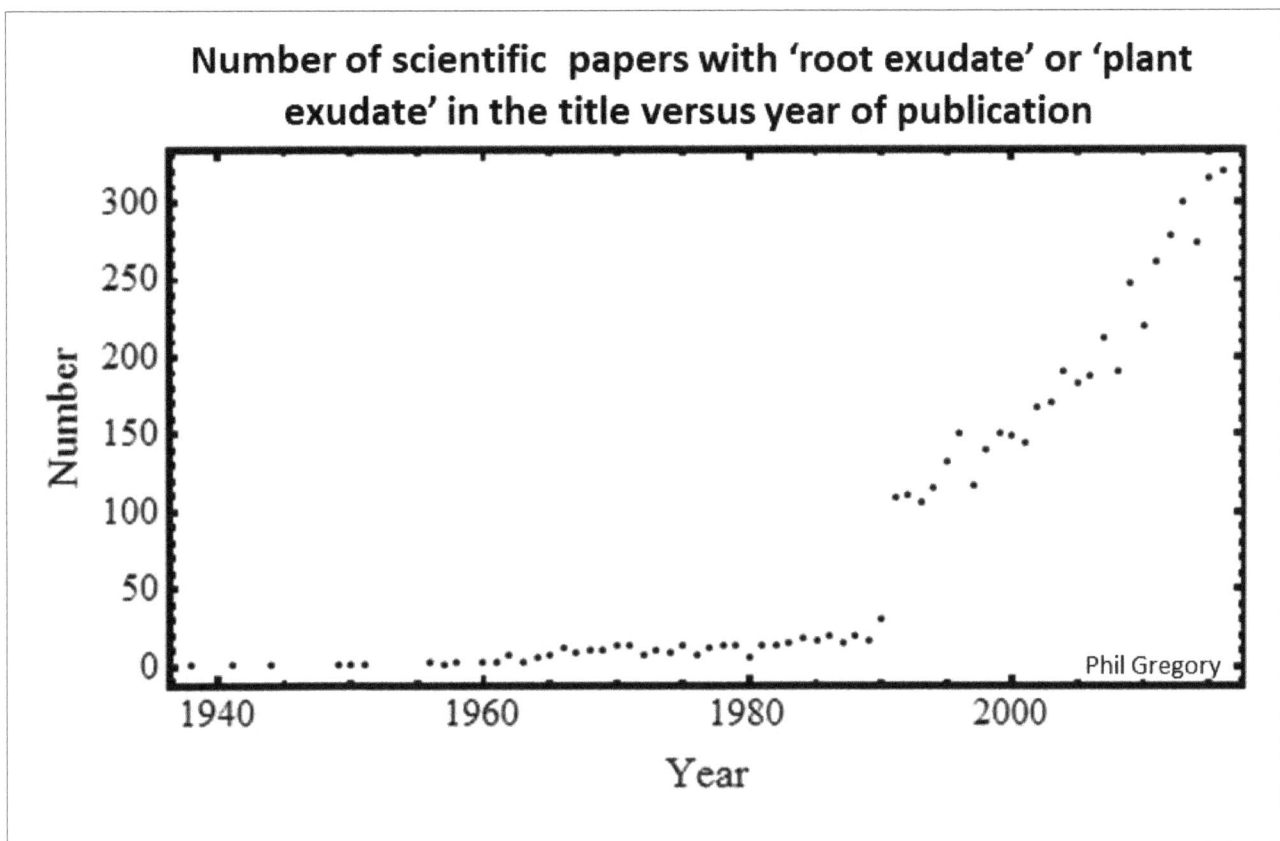

Figure 2. One indicator of the recent revolution in soil biology is the sudden jump in the number of scientific papers on exudates that occured in 1991. A key paper that is considered a prime catalyst of this revolution is, *Interactions of Bacteria, Fungi, and their Nematode Grazers: Effects on Nutrient Cycling and Plant Growth*, by Russell E. Ingham, J. A. Trofymow, Elaine R. Ingham, and David C. Coleman, Ecological Monograph*s,* 1985, Vol. 55, No. 1, pp. 119-140.

Figure 3. Plants release up to 40% of the sugars, produced by photosynthesis, as root exudates to attract and power soil microbes. According to Dr. Elaine Ingham, these root exudates are like cakes and cookies to the microbes.

In return, the microbes provide all the extra nutrients plants require by:
- mining the rocks, sand, silt, & clay,
- recycling dead plant and animal matter,
- fixing nitrogen from the atmosphere.

In healthy soil conditions, leaf surfaces are covered by microbes held to the plant by the strong biotic glues. That protective layer is one of nature's ways of achieving disease suppression.

Image: Exposed roots on a mango tree by Aaron Escobar (CC By 2.0)
https://upload.wikimedia.org/wikipedia/commons/2/29/Exposed_mango_tree_roots.jpg
Modified in 2021 by Phil Gregory to illustrate plant exudates

Chapter summary:
- Plants capture sunlight energy and store it in the carbon bonds of sugars and carbohydrates they produce during photosynthesis.
- Plants release up to 40% of these carbon compounds as root exudates to attract and feed soil microbes in return for many other nutrients. This is nature's barter system.
- Microbes obtain the additional nutrients plants require by recycling dead plant and animal matter and by mining the rocks, sand, silt, and clay.
- Chemical-intensive agriculture shuts down this barter system and leads to a dependence on pesticides.
- A massive increase in the use of chemical herbicides that are chelators makes many metals unavailable to the growing plant leading to plant deficiencies.

6. Our choice: put more carbon into the atmosphere or sequester it in the soil

In the 1990s, Dr. Don Reicosky and colleagues of the USDA made a surprising discovery (see Figure 4). They found that 15 times more carbon dioxide was released during a 24 hour period after the soil was plowed to a depth of 28 cm, than from the neighbouring untilled soil[30]. When the sample period after plowing was extended to 21 days the emissions were still 10 times higher. Carbon dioxide (CO_2) gas is invisible and odourless so it came as a real shock to find out how much extra greenhouse gas was produced by plowing and how unsustainable plow-based agriculture is. Apparently, plowing wakes up opportunistic r-strategist soil bacteria which rapidly increase in numbers. They convert stored soil organic carbon into carbon dioxide that escapes into the atmosphere.

On the flip side, another huge role that microbes can play is to actually sequester carbon in the soil, a powerful way to draw down carbon from the atmosphere. Even if we stopped all human emissions today the climate would still continue to warm. We have to actually remove some of the excess carbon from the atmosphere. It now appears that we can efficiently return carbon to the soil[31][32][34][37][38][39][40][41][42][43][44] by utilizing nature's biological solutions. If we change the way we do agriculture, Nature will do a lot of the work for us through the actions of the soil biology which includes a vast collection of microbes.

It is expected that as the soil takes up more carbon from the atmosphere the ocean will release some of its excess carbon to the atmosphere in a rebalancing, so it may take a long time before atmospheric carbon actually declines. Of course, this rebalancing will be good for life in the oceans and the extra carbon in the soil will increase soil fertility and rehydrate the soil allowing for more plant growth, more evapotranspiration and more cooling. Evapotranspiration means the process by which water is transferred from the land to the atmosphere by evaporation from the soil and other surfaces and by transpiration from plants.

Chapter summary:
- Plowing and tillage greatly enhances the release of soil carbon as CO_2, an atmospheric greenhouse gas.
- We can efficiently sequester carbon in the soil using nature's biological solutions. This will rehydrate the soil allowing for more plant growth, more evapotranspiration, and more cooling.

6. Our choice: put more carbon into the atmosphere or sequester it in the soil

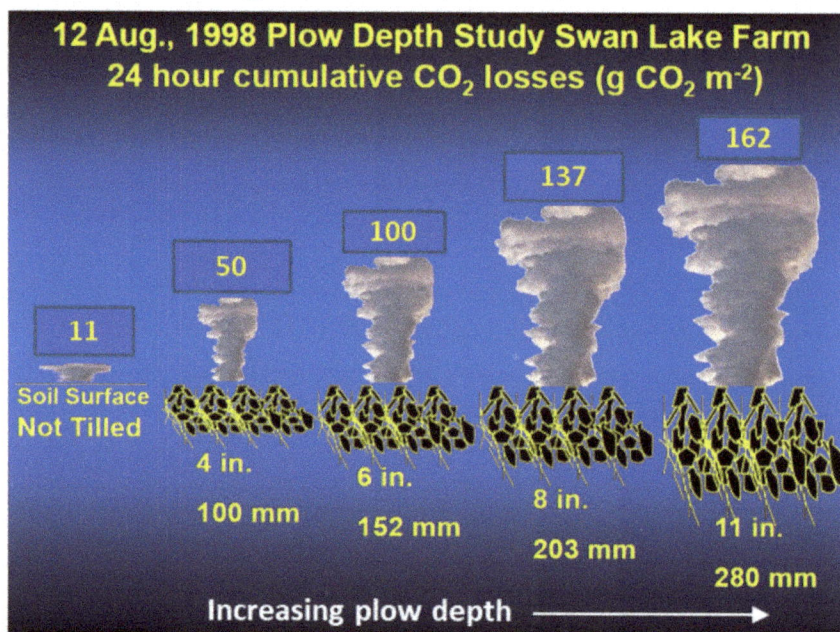

Figure 4. To measure the CO_2 greenhouse gas emissions, MR. GEM is parked over a patch of earth and records the CO_2 given off in a given time period. (images provided by Dr. Don Reicosky, USDA Agricultural Research Services, Morris, MN)

7. The soil carbon sponge

Didi Pershouse, author of *Understanding Soil Health and Watershed Function*, uses a useful analogy between soil that has been turned to dirt and a pile of baking flour [45]. If you simulate rainfall onto a pile of flour it erodes off the surface layer. Very little water penetrates into the pile because the smaller flour grains fill the surface voids between the larger grains and just below the surface the pile is dry. But when you add biology to the flour in the form of yeast, a fungus, and make bread, the flour now has structure and acts like a sponge. When you rain water on the bread it soaks in (infiltrates) and is stored, and there is no erosion. A display of the flour and bread experiment is shown in Figure 5. The microbes in healthy soil provide a similar sponge-like structure which is referred to as the soil carbon sponge.

Through plowing and chemicals we destroy the biology and soil structure. This results in more frequent floods, much greater sensitivity to drought, and landscapes more prone to fire. The way we have been practicing agriculture has reduced soil organic carbon from about 4% to 5% in native soils to 1% or less in most agricultural soils today. Soil organic matter is approximately 58% carbon. The organic matter is composed of living roots and organisms (5%), dead roots and organisms including biotic glues (less than 10%), actively decomposing roots and organisms (25%), and more stable organic matter called humus 50%-80%.

We need to rehydrate our soils by building up the soil carbon sponge through a healthy soil biology. Even a 1% increase in soil organic carbon translates to an increase in the water holding capacity of approximately 20,000 US gallons per acre[46]. This action can help compensate for greater rainfall variability associated with climate change.

That soil sponge structure is created by the microbes building their underground cities. We need more civil engineers and landscapers who are trained to appreciate the important role biology can play in flood, drought, and fire protection. The Australian soil ecologist, Dr. Christine Jones, provides an excellent explanation of how soil biology can restore soil carbon and rapidly build topsoil[47].

Chapter summary:
- Biology can transform a pile of flour into bread and a pile of dirt into soil, a water holding carbon sponge.

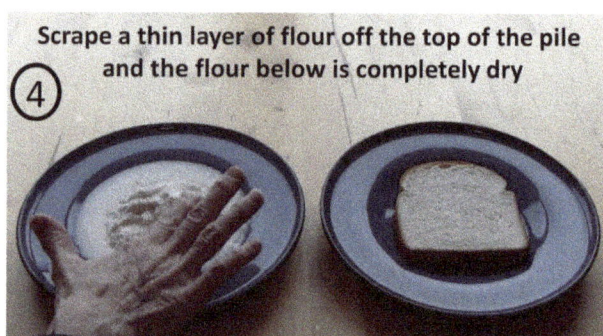

Figure 5. A simple flour and bread exercise to help convey the relationship between soil and water.

8. Saskatchewan no-till farmers

After learning about the extra CO_2 released when the soil is plowed or tilled, I asked Dr. Reicosky if he was aware of significant no-till activity happening on Canadian farms. He drew my attention to a 2016 article about a group of Saskatchewan farmers who had been sequestering soil carbon for 15 years through no-till farming[48]. Because of extreme water shortages more than 60% of Saskatchewan crop farmers made the change to no-till farming. The latest Statistics Canada figure[49] lists Saskatchewan's no-till or zero till seeding at 74% in 2016. The average for Canada as a whole was 59%.

Sakatchewan went from one commercial crop every two years to one every year and returned to making a profit. According to a Scientific American report[50], no-till practices decrease the fuel expense by 50% to 80% and the labour by 30% to 50% but often rely heavily on herbicides to control weeds.

The Saskatchewan Soil Conservation Association (SSCA) published a position paper discussing the results of the Prairie Soil Carbon Balance (PSCB) project, which measured changes in soil organic carbon in 137 Saskatchewan field sites under direct seeding management over a period of 15 years. They demonstrated that a significant amount of CO_2, averaging 0.94 tons of CO_2 per hectare per year, was sequestered in 27,964,691 acres (2016 Census) of no-till crop management in the province[51]. This is equivalent to removing 2 million cars from the roads (more than 2.5 times the number of light motor vehicles registered in all of Saskatchewan in 2016).

I had several opportunities to talk with a representative of the SSCA to learn more about their no-till experience which I briefly summarize in the following panel.

> **From a conversation with a Saskatchewan Soil Conservation Association representative**
>
> 1. It took about three years from commencing no-till before they were able to move from one cash crop every two years to one every year.
> 2. I asked how wide spread no-till practices were in other Canadian provinces? Apparently the practice is generally growing[49] but is still not widespread because farmers in many other provinces don't suffer from the same extreme water shortages.
> 3. I learned that farming is a high-risk occupation and that in general if a farmer is making a profit or even breaking even they are very reluctant to make any change.
> 4. I asked if Saskatchewan farmers were also experimenting with moving away from the use of fossil-fuel based fertilizers and pesticides to follow the example of pioneering regenerative agriculture farmers like Gabe Brown of North Dakota, who uses diverse cover crops and livestock integration. I was given three responses to this question:
>   - The first was currently a **no** for the reason given in item (3) above.
>   - The second was that it stands to reason that if they export their crops to another country they are exporting their soil nutrients so they must add these nutrients back with chemicals.
>   - The third, Gabe Brown was the keynote speaker at their upcoming 2017 conference.

I was intrigued by the issue of exporting soil nutrients and I will explore that question in a later chapter. Before that we have the more basic question of what are plants made of?

9. What are plants made of?

One of the first answers came from a willow tree experiment carried out by Jean Baptista van Helmont (1577-1644). At the time it was not considered safe to ask questions about God's creations and Helmont had already been arrested by agents of the Spanish Inquisition for the crime of studying plants and other phenomena. The prevailing theory was that plants grew by eating soil, and van Helmont devised a clever test of this idea. He planted a small willow tree in a pot of dry soil weighing 90 kg. He figured that if the tree ate the soil, then the weight of the soil should decrease over time. For five years he watered and took care of the willow. It grew from 2.2 kilograms to 77 kilograms (169 pounds), while the dry weight of the soil had lost only about 57 grams (2 ounces)!

One surprising conclusion from this experiment is that the soil minerals only contributed a tiny percentage (of order 0.08%) to the weight of the tree. Helmont's willow experiment, which is probably the very first scientific experiment in plant nutrition, was published after his death, in Ortus Medicinae (1648)[52].

About 200 years later, a Swiss chemist, Nicolas-Théodore de Saussure (1767-1845), proved Steven Hales's theory that plants absorb water and carbon dioxide in the presence of sunlight and increase in weight. Saussure was thus one of the major founders of the study of photosynthesis. He further demonstrated that plants are dependent upon the absorption of nitrogen from soil. Subsequently, we learned that the soil nitrogen is derived from atmospheric molecular nitrogen that has been converted to plant available nitrogen compounds by soil microbes (bacteria and archaea) in a process called nitrogen fixation. Biological nitrogen fixation is one of the most important ecological processes on Earth.

So, it turns out that trees and plants are mainly made from the constituents of air, like carbon dioxide, nitrogen (through the action of soil microbes), and water that comes down from the atmosphere in the form of rain. There is an interesting BBC video clip of the Nobel Prize winning physicist, Richard Feynman, speaking about this topic[53].

What we have learned is the vast majority of the weight of a tree is made of elements that can be obtained from the air and water that comes down as rain. This appears to be true of all plants. So it is only a small but very essential soil mineral component that might become depleted when we export crops from the farm. In the next chapter we will take a closer look at the elemental makeup of a corn plant.

Chapter summary:
- Plants are mainly made from the constituents of air like CO_2, water that comes down as rain, and nitrogen. The CO_2 and water are converted to sugars by photosynthesis. Soil microbes are responsible for converting atmospheric nitrogen to plant available nitrogen compounds.
- Other elements derived from soil minerals amount to much less than 1%.

10. Exporting soil nutrients

When we export our harvest from the farm, are we exporting some of the soil minerals faster than they can be replaced naturally? Many farmers and agronomists argue that we need to replace at least some key nutrients like nitrogen (N), phosphorus (P) and potassium (K) by adding chemicals. Why nitrogen? There is plenty of that in air which can be converted to plant available nitrogen compounds by nitrogen fixing microbes.

The problem is that we humans have badly degraded our soil through plowing as mentioned earlier. This slices and dices the soil carbon sponge structure built by microbes. As a consequence of the soil degradation we have badly impacted the water cycle and microbe mineral extraction, turning living soil into dirt. We learned from the green revolution that it is possible to grow plants in dirt, in the short term, if we add nitrogen fertilizer together with phosphorus, potassium and supporting pesticides. In the following chapter on "The browning of the Green Revolution," we will see some of the downside of the use of synthetic nitrogen fertilizer.

The last 30 years has seen a revolution in our understanding of soil biology. We now understand how nature's barter system operates between plants and soil microbes to provide plants with all the additional soil nutrients plants require. We now know how to restore the soil biology and rebuild the soil carbon sponge, key aspects of regenerative agriculture that reverse soil degradation. Regenerative agriculture is discussed more fully in later chapters. Our new knowledge allows us to work with nature to grow our food in ways that are congruent with our need to get off of fossil fuels.

In the panel that follows, I estimate the rate at which essential soil minerals like phosphorus, potassium, and magnesium need to be supplied by soil microbe mining operations, to keep up with the rate at which these minerals are exported in a typical corn crop. Feel free to skip over this panel if you are less interested in the details of this calculation.

> **Details:** For a corn plant the moisture content is about 71%[54]. Approximately 99% of the dry weight of the whole corn plant (after removing the moisture content) is composed of 45% carbon, 46% oxygen, 6.4% hydrogen, and 1.4% nitrogen, all things sourced from the air[55]. That only leaves 1.2% of the plant's dry weight or 0.35% of the plant's full wet weight required to come from the soil in the form of about 17 other elements, mainly in trace amounts.
>
> I will focus on the amount of phosphorus (P) that is exported in corn kernels because the plant's demand for P is the highest relative to its abundance in the soil. Below is the calculation of an estimate of the depth of soil that needs to be created to supply the exported phosphorus each year.
> - In the U.S. the average corn yield is 168.4 bushels of corn per acre[56].
> - One bushel corresponds to 56 lb of corn kernels so what is being exported is 9430 lb/acre = 4.72 tons/acre = 10.6 metric tonnes/hectare (ha).
> - Phosphorus amounts to 0.34% of dry weight[55] or 0.099% of wet weight of a corn kernel. Therefore the fraction of corn weight arising from phosphorus = 0.099% /100 = 0.00099
> - Weight of phosphorus exported in the corn crop = 0.0047 tons/acre = 0.0105 tonnes/ha
> - Median abundance of phosphorus[57] in soil is 800 mg/kg = 0.0008 kg of P/kg of soil
> - This translates to a weight of soil = 13.1 tonnes/ha = 5.85 tons/acre, to provide the exported P.
> - A 1" layer (25.4 mm) of soil over one acre weighs about 150 tons[58]
> - Therefore, the depth of soil to provide the exported phosphorus = 1 mm/year
>
> Repeating this calculation for potassium and magnesium, the estimates of the depth of soil required each year are much less, coming in at 0.07 mm/year and 0.094 mm/year, respectively.

The conclusion of the above panel is that to keep up with the rate at which phosphorus is exported in a typical corn crop, we need to create new soil at the rate of 1mm/year. The number often used is that it takes 500 to 1000 years to produce 3 cm of soil by physical and chemical weathering of rocks which translates to a rate of 0.06 mm/year to 0.03 mm/year. So can a healthy population of soil microbes, including mycorrhizal fungi, mine and recycle an amount of phosphorus each year equivalent to that contained in 1 mm of new soil? **Unfortunately, my search of the scientific literature did not yield a definitive answer to the rate of biological weathering.** Just recently another mechanism was identified that enables bacteria to degrade bedrock, jump-starting a process that creates the mineral portion of soil but no rate for this process is presently available[59].

What we learned from our calculation above is that to sustainably export corn we need to be creating the equivalent of 1 mm of new soil each year to provide for the phosphorus exported in the corn kernels. Recall, from Chapter 4, that the median soil erosion rate is 1.5 mm/year[18] which means we are losing soil faster than we would like to be creating it. Again, this just highlights why we need to stop agricultural practices that are causing this erosion, like plowing and tilling.

Some inferences can be made about biological soil nutrient creation from pioneering regenerative agriculture farmers like Dr. Allen Williams, Gabe Brown, Joel Salatin, and many others that are documented in Gabe Brown's book, *Dirt to Soil*[34]. Australian farmer and author, Dr. Charles Massy, provides other fascinating examples of successful transitions to regenerative agriculture in his ground breaking book, *Call of the Reed Warbler*[35]. I was particularly interested to learn how big an impact the works of Dr. Elaine Ingham and Allan

Savory (see later chapters) were having in Australia. The two books by Gabe Brown and Charles Massy provide a clear vision of a sustainable future for our food supply.

In 2003, Dr. Kris Nichols, a soil microbiologist at the USDA Agricultural Research Services in Manden, North Dakota, visited Gabe Brown's farm. Her advice to Gabe was that his soils would never be sustainable until he stopped using synthetic fertilizers[34]. They interrupt the relationship between plant roots and microbes and only supply a limited range of nutrients. The fertilizer provides the plants with certain key nutrients, so the plants stop using the barter system to trade carbon for nutrients from the microbes. The plants keep the carbon for themselves which means the microbes don't get enough carbon exudates and their population suffers. Mycorrhizal fungi efficiently acquire a wide range of minerals for plants in exchange for carbon from the plants, but if the fungi do not receive carbon, they can't acquire these minerals. To find out if his soils were already healthy enough that he didn't need use synthetic fertilizers, Gabe carried out four years of split trials comparing with and without synthetic fertilizer, starting in 2004. For four years in a row, the crop yields of the unfertilized half were equal to or greater than for the fertilized half and Gabe also noticed a dramatic improvement in the soil health once he removed synthetic fertilizers. Among other things, the soil particles were more aggregated, meaning they were held together by biotic glues and fungal strands. Also, water infiltration was significantly improved.

Gabe Brown's book, *Dirt to Soil*, documents other tests of his soils carried out by Dr. Rick Haney at the USDA Agricultural Research Services in Temple, Texas. The testing compared the results of Gabe's regenerative agriculture management to three other neighbouring farms with the same soil type that used very different management styles. The others ranged from organic farming using tillage, to no-till, high diversity crops using synthetic fertilizers, pesticides, herbicides, and fungicides. Gabe stopped using synthetic fertilizers in 2007 and ceased using pesticides or fungicides before 2000. In 2015, the soil on Gabe's farm was found to have at least 7.5 times the levels of nitrogen, 3.9 times the level of phosphorus, 8.8 times the level of potassium, and 4.1 times the amount of organic matter. While the infiltration rates on the other three farms were between 0.5 inches to 0.7 inches per hour, Gabe's infiltration rate was 30+ inches per hour. All infiltration rates were measured on the same day. Gabe experienced some of his biggest improvements after integrating livestock into his operation.

One very profitable regenerative farming example is Joel Salatin's Polyface Farm in Virginia. He turned a $300-a-month subsistence farm into $2-million-a-year organic business[36]. The 2000 acre farm supports 20 full-time salaries and offers a paying internship program for young, would-be producers. According to their website, they haven't bought a bag of chemical fertilizer in half a century, never planted a seed, owned a plow, disk or silo. "We practice mob stocking herbivorous solar conversion lignified carbon sequestration fertilization with the cattle. The Eggmobiles follow them, mimicking egrets on the rhinos' nose. The laying hens scratch through the dung, eat out the fly larvae, scatter the nutrients into the soil, and give thousands of dollars worth of eggs as a byproduct of pasture sanitation. Pastured broilers in floorless pasture schooners move every day to a fresh paddock salad bar. Pigs aerate compost and finish on acorns in forest glens. It's all a symbiotic, multi-speciated synergistic relationship-dense production model that yields far more per acre than industrial models." Joel believes in diversification and direct marketing, avoiding excessive debt by purchasing used equipment, starting new operations on a small scale, keeping infrastructure portable, renting rather than buying land.

Rudolf Steiner, the father of biodynamic farming where no inputs external to the farm are used, believed that without bovine manure it is impossible to maintain healthy soil. He believed that animals, crops, and soil are part of a single system.

There is an often heard phrase that "Soil is the Foundation of Life." We tend to view soil as the base of the food chain. The Australian ecologist, Christine Jones, thinks we have perhaps got this backward and that photosynthesis, meaning 'life from light', is the true base. Green plants, sunlight, CO_2, and water, produce sugars that store sunlight. These carbohydrate exudates, that are released by the plant roots, power the soil microbes that recycle organic matter, extract new soil minerals from the rocks, sand, silt, and clay[63][47], and build soil. I will return to the subject of microbes and soil building in a later chapter.

More on P availablity: According to Christine Jones, it is widely recognized that only 10%-15% of P fertilizer is taken up by crops and pastures in the year of application[64]. Phosphorus is a highly reactive element. Approximately 80% of any free phosphorus floating around in the soil forms a chemical bond with another element like iron or aluminum or calcium, making it unavailable to plants. But certain bacteria produce an enzyme called phosphatase that can break that bond and release the phosphorus. Once released, the phosphorus still has to be transported back to the plant, which is where mycorrhizal fungi come in. They form networks between plants and colonies of soil bacteria. According to Christine Jones, if P fertilizer has been applied for the previous 10 years, there will be sufficient P for the next 100 years, irrespective of how much was in the soil beforehand. Rather than apply more P, it is more economical to activate soil microbes and their mycorrhizal networks in order to access the P already there.

Earlier I mentioned the importance of mycorrhizal fungi in extending plant root area for extracting nutrients. It turns out the enzymes the fungi secrete also play an important role for increasing the availability of soil P [14]. Mycorrhizal fungi transport a wide variety of other nutrients, including nitrogen, sulphur, potassium, calcium, magnesium, iron, water, and essential trace elements such as zinc, boron, manganese and copper. Fungi can absorb nutrients across their entire bodies and not just the tips. Mycorrhizal fungi are both the highway and the internet of the soil.

Mycorrhizae abundance can be significantly improved through cover crops, diversity, and appropriate grazing management. The cover crops extend the period of time that plants are pumping out liquid carbon exudates (sugars) to power the recycling and mining operations carried out by the microbes.

Chapter summary:
- Only about 0.3% of a plant's weight comes from soil minerals, the rest comes from the air, including water that arrives as rain and atmospheric nitrogen fixed by soil bacteria.
- To supply the phosphorus, that gets exported in a typical corn harvest, requires a soil formation rate of about 1 mm per year which is considerably higher than what can be achieved by the physical and chemical weathering of rock. The soil formation rate requirements for potassium and magnesium are much less than for phosphorus.
- Currently, the scientific literature does not provide a good answer as to how quickly the soil mineral component can be biologically mined from the rock, sand, silt, and clay.
- However, there are a growing number of successful regenerative agriculture farmers who have not used plowing or synthetic fertilizers for a decade or more and their soil fertility is increasing.

11. Nutritional declines in foods

A widely quoted paper[26] on the topic of changing food nutrition was published in the peer reviewed journal Nutritional Science in 2003 by David Thomas. The Journal provided the following information about the author. David Thomas is a geologist with an M.Sc. in Mineral Exploration who worked for nine years in copper, cobalt, lead, zinc, gold and uranium exploration and mining and is a Fellow of the Geological Society. Subsequently he retrained in the field of Chiropractic and later in Nutrition, being a founding Member of the Register of Nutritional Therapists. In addition to his practice, David Thomas is the UK Distributor of a multi trace element supplement.

The paper presents Thomas's analysis of data from McCance and Widdowson's epic work, *The Composition of Foods* (a reference manual republished and updated by Government biochemists every few years). Here is a quote from the abstract of his paper. "In 1927 a study at King's College, University of London, of the chemical composition of foods was initiated by Dr McCance to assist with diabetic dietary guidance. The study evolved and was then broadened to determine all the important organic and mineral constituents of foods, it was financed by the Medical Research Council and eventually published in 1940. Over the next 51 years subsequent editions reflected changing national dietary habits and food laws as well as advances in analytical procedures. The most recent (5th Edition) published in 1991 has comprehensively analyzed 14 different categories of foods and beverages. In order to provide some insight into any variation in the quality of the foods available to us as a nation between 1940 and 1991 it was possible to compare and contrast the mineral content of 27 varieties of vegetable, 17 varieties of fruit, 10 cuts of meat and some milk and cheese products. The results demonstrate that there has been a significant loss of minerals and trace elements in these foods over that period of time."

During that 51-year period, for each mineral, the cumulative total content in 27 vegetables showed a loss amounting to 76% in copper, 46% in calcium, 24% in magnesium, 16% in potassium and 9% in phosphorus. For ten meats, the mineral content for the cumulative totals showed a loss of 54% in iron, 41% in calcium, 24% in copper, 16% in potassium, and 10% in magnesium. For 17 fruits the average minerals losses ranged from 2% to 27%. Thomas reports that the most dramatic losses relate to copper in vegetables which between 1940 and 1991 declined by 76%, and between 1978 and 1991 zinc declined by 59%. Thomas proposed a lengthy list of factors to do with how we grow and process our foods that contributed to changes in food nutrition over time. Although the McCance and Widdowson data is probably the best available for such an historical study, it is difficult to assess Thomas's conclusions in this paper and his subsequent 2007 paper [252], because he does not quote uncertainties in the values or compute the statistical significance of these changes. On the basis of his analysis I can only conclude that he found evidence for an apparent change in the mineral content over time.

Many researchers have questioned the accuracy of studies spanning such a long time in part because of improvements in methodology. According to Professor Michael Crawford of London Metropolitan University, the traditional methods used by physical chemists are still the most accurate on the planet for measuring weights and components and that what has changed is the computerized automation of the measurements not the accuracy[254].

In 2004, Professor Donald R. Davis and collaborators[253] carried out a rigorous statistical comparison of USDA Food Composition data from 1950 and 1999, for water, energy, protein, fat, carbohydrate, ash, calcium, phosphorus, iron, vitamin A, thiamin, riboflavin, niacin, and ascorbic acid in 43 garden crops. They found that changes for individual foods could not be assessed reliably due to large uncertainties in the mineral nutrient content data, but grouped together, statistically significant decreases from 1950 to 1999 were seen for 6 nutrients (protein, calcium, phosphorus, iron, riboflavin, and ascorbic acid). Calcium declined by 16%, phosphorus by 9%, and iron by 15%. No statistically reliable changes were found for 7 other nutrients.

In 2009, Professor Davis published[255] an informative review paper titled *Declining Fruits and Vegetable Nutrient Composition: What is the Evidence*? The results discussed here are from his review paper and a 2019 lecture he gave that is available on YouTube[256]. One method already mentioned are recent studies of historical food composition data. Of particular interest is a comparison of the best statistical analyses, including his own, for U.K. data for vegetables and fruits. The conclusion of that work is the finding that for 11 nutrients the declines are statistically significant. The strongest evidence for declines occurs for minerals in vegetables, especially copper, magnesium, and calcium, with declines of 80%, 22%, 17%, respectively. Among the 33 nutrients examined, none exhibited a statistically significant increase.

A second line of evidence comes from early studies of fertilization which found that increasing the amount of fertilizer increased the dry plant weight (yield) but the increase in nutrient absorption was often less, leading to a dilution of these nutrients in the overall dry plant weight (after water removed). In one experiment with red raspberries published in 1979, three different phosphorus (P) fertilization amounts were used of 0 parts per million (ppm), 22 ppm and 44 ppm. The relative dry weight of the plant went from 1.0 for 0 ppm P, to 1.4 for 22 ppm P, to 2.2 for 44 ppm P. Increasing P fertilization caused a decline in eight minerals with the biggest decline for copper coming in at 58%. This type of "dilution effect" is now referred to as environmental dilution.

Another example of environmental dilution is nutrient decline caused by increased levels of the atmospheric greenhouse gas CO_2. Higher levels of atmospheric CO_2 result in increased photosynthesis production of carbohydrates which are by far the biggest contribution to dry plant weight. One study on rice carried out in Japan simulated the projected atmospheric CO_2 for the year 2100. The results showed a significant decline in 7 nutrients but an increase in vitamin E. The biggest decline of 30% occurred for folate (vitamin B9).

There is another different nutrient dilution effect that is caused by genetic changes in plants through breeding and hybridization programs. Plant hybridization is the process of crossbreeding between genetically dissimilar parents to produce a hybrid. It frequently results in offspring with reduced fertility allowing the production of seedless varieties. Frequently these breeding programs are designed to improve traits (size, growth rate, pest resistance) other than nutrition. A good example of the resulting "genetic dilution" is the case of the super food broccoli. According to Davis, 14 varieties of broccoli were introduced from 1954 to 2004 mainly to increase the head weight. Experimenters reported a strong inverse correlation between head weight and mineral concentration with large declines in copper, zinc, iron and manganese[256]. These experiments are done with a collection of different breeds of a single food grown side-by-side for purposes of comparing their nutrient content. Side-by-side experiments eliminate the need for historical data that requires averaging over large numbers of foods.

According to Davis, broccoli is just one example of many cases where nutrient losses have been measured in breeding programs for high yield in many varieties of wheat, maize, potatoes, and rice. In these side-by-side experiments, unwelcome, unintended declines in nutrients rarely attract study or mention by researchers thus many questions remain unanswered. He highlights the need for further studies to assess the generality of dilution effects among foods and to greatly expand the numbers of nutrients and important plant produced chemicals compounds (phytochemicals). He recommends side-by-side comparisons studies in multiple environments to provide rigorous answers to the many remaining uncertainties.

In a recent side-by-side comparison study, Sovan Debnath and eight other scientists found a downward trend in the grain density of zinc and iron in high-yielding cultivars of rice and wheat released in succeeding decades since the beginning of green revolution in India[263]. Rice produced from cultivars released in the 1960s were 24.0% and 28.0% higher in zinc and iron, respectively, compared to cultivars from the 2000s. For wheat, the concentrations of zinc and iron were 29% and 19.5% higher, respectively, in the 1960s cultivars compared to 2010 cultivars. They attribute this to the inefficiency of modern-bred cultivars of rice and wheat to sequester those essential nutrients.

Related to this topic is the observation that many cultivars, that have been produced by selective breeding, have lost some ability to interact with soil microbes that provide necessary nutrients. Stephanie Porter and Joel Sachs at Washington State University, Vancouver, reviewed 120 studies of microbial symbiosis in plants and concluded that many types of domesticated plants show a degraded capacity to form symbiotic communities with soil microbes[260]. Benefits like phosphorus provisioning by mycorrhizae or symbiotic nitrogen fixation by rhizobia, can be inhibited or negated under fertilization if these nutrients are freely available to plants in the soil. This has made some domesticated plants more dependent on fertilizer, one of the world's largest sources of nitrogen and phosphorus pollution and a product that consumes fossil fuels to produce. Apparently, the solution is to breed the domesticated cultivar with a wild relative that has not lost those traits. Future breeding trait studies need to examine the cultivar's ability to form symbiotic communities with soil microbes. These studies should recognize the important interactions between plants, bacteria, fungi, and a hierarchy of predators that make up the soil food web[4][5].

There is now mounting evidence for nutrient declines in our basic foods stocks which are usually in the range 5% to 35%, rarely 75% except in the case of copper. Davis points out that these declines are smaller than the nutrient losses in refined and processed foods so vegetables and fruits still remain our main source of nutrients.

Chapter summary:
- There is mounting evidence for nutrient declines in our basic food stocks which are usually in the range 5% to 35%, rarely 75% in the case of copper.
- Nutrient dilution can result from using fertilizers.
- Nutrient dilution can result from selective breeding for high yield.
- Nutrient dilution can result from hybridization.
- Nutrient dilution can result from increased atmospheric CO_2.
- These declines are not as large as nutrient losses in refined and processed foods.
- Vegetables and fruits still remain our main source of nutrients.
- Breeding trait studies should also examine the new plant's ability to form symbiotic communities with soil microbes, like nitrogen fixing rhizobia.

12. The browning of the green revolution

Since the Green Revolution of the 1960s, substantial increases in cereal production have occurred as a result of the introduction of high-yielding varieties of grains and heavy inputs of fossil-fuel based synthetic fertilizers and pesticides. This approach was directed toward maximizing grain yield, without regard to long-term impacts on the soil resource that is crucial for sustainable cereal production.

In 2009, three University of Illinois scientists, R. L. Mulvaney, S. A. Khan, and T. R. Ellsworth, published a paper entitled *Synthetic Nitrogen Fertilizers Deplete Soil Nitrogen: A Global Dilemma for Sustainable Cereal Production*[65][66][67][68]. The University of Illinois runs the Morrow Plots, the oldest continuous agricultural research fields in the United States that have been in operation since 1876.

For 40 to 50 years the Morrow Plots have been used to document the impact of using high-input agriculture on crop yields and soil properties. Since the level of nitrogen applied was greater than the amount being removed in the crop, the expectation was that it would accumulate in the soil. Nitrogen inputs, ranging from 1.8 to 5.6 tons per acre for 51 years, provided at least 60% more nitrogen than the cumulative amount removed in corn grain, yet there was a net decline of 624 to 1606 pounds per acre in total soil nitrogen[67]. They arrive at a similar conclusion from an examination of nitrogen data from many other published cropping experiments that document nitrogen changes over time for fertilized soils encompassing a global range of climates, soils, cropping systems, and management practices.

The problem is that synthetic nitrogen fertilizers stimulate microbial organic carbon decomposition, promoting the loss of crop residues and also indigenous organic matter that serves as the major reservoir of soil nitrogen. Manure is a mix of slow-release organic nitrogen and organic matter while synthetic nitrogen fertilizer is pure, readily available nitrogen. A large amount of the readily available nitrogen leaches away, fouling ground water in the form of nitrates, and entering the atmosphere as nitrous oxide (N_2O), a greenhouse gas with some 300 times the heat-trapping power of carbon dioxide. For every available nitrogen atom it consumes, a bacterium needs five carbon atoms which they grab from the soil organic matter which is approximately 58% carbon. This results in a decrease in the all important soil organic carbon.

Higher nitrogen rates may offer temporary relief, but the long-term consequences will be a further decline in soil productivity that increases the need for synthetic nitrogen fertilization, intensifies food insecurity, and exacerbates environmental degradation. Their conclusions corroborate the view elaborated long ago by White (1927)[69] and Albrecht (1938)[70] that nitrogen fertilizer depletes soil organic matter by promoting microbial soil carbon consumption and a loss of stored soil nitrogen. Another recent study provides solid evidence that biological nitrogen fixation will be largely suppressed in an increasingly fertilized world, with implications for soil biodiversity and ecosystem functions[71].

A number of more recent papers show evidence of crop yield stagnation in the world's major cereal crops, including maize, rice and wheat that appears to have started in 1990s[72][73][74][75]. For example, Australia's wheat yields more than tripled during the first 90 years of the 20th century but have stalled since 1990[75]. In Asia, the major rice growing continent, yields are stagnating in China, India and Indonesia across 79%, 36% and 81% of rice-growing areas, respectively[72]. In some cases the reductions are attributed to reduced rainfall and rising temperatures, both trends expected from climate change. In following chapters, I will look at how changes to the way we practice agriculture can make a positive difference.

Chapter summary:
- In the long term, synthetic nitrogen fertilizer depletes soil organic matter and stored nitrogen, leading to crop yield stagnation.

13. Nature's complexity is amazing

In the past 20 to 30 years a new paradigm has emerged in our understanding of how to do agriculture in a sustainable way and it is all about biology. Much of the new findings have to do with the soil's microscopic organisms. It turns out we have also learned a lot recently about how nature evolved large herbivores and their predators to build and maintain healthy grasslands.

For 10,000 years we did not understand the complex web of interactions between grassland, herbivores, and their ferocious predators. In our ignorance we have turned much of the world's grasslands to desserts. As the African ecologist Allan Savory has shown, it is not the numbers of animals that is responsible for desertification, it is our failure to manage plant recovery time that was achieved by nature through a balance of herbivores and predators[76][77]. The panel below provides a simple way to understand this from the sigmoid growth curve of grass.

After being grazed to the ground level the growth of the grass is initially slow in the baby phase, gradually accelerating to enter the rapidly growing teenage phase, after which the growth slows down and stops in the final phase.

When herbivores are left to graze in a pasture for a season, the grass spends most of the time growing slowly in the baby phase. Repeated cropping at that stage fails to recharge the roots adequately through photosynthesis. In drought prone areas this repeated cropping can kill the plant and lead to desertification.

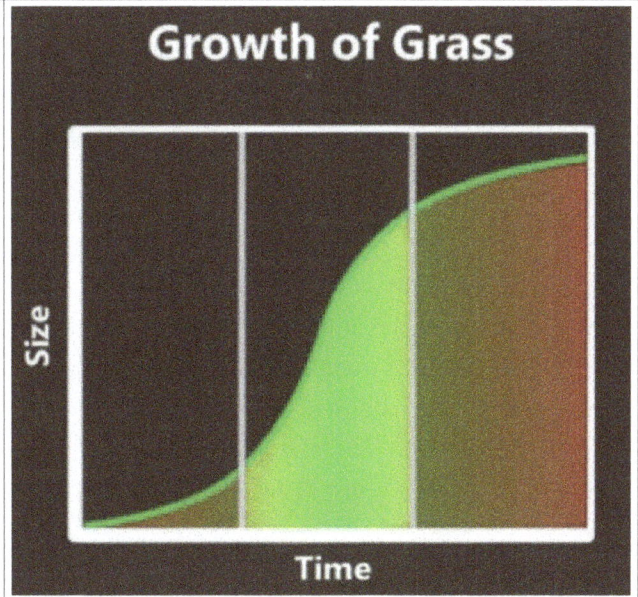

Figure 6. The sigmoid growth curve of grass divided into three time phases. Image credit Jimi Eisenstein[78].

In nature, the predators keep the herbivores bunched together and moving so they don't get to eat the grass a second time as it starts to regrow and recharge the plant roots. By the time they return from their migration, the grass is fully grown and needs to be eaten. Figure 7 shows some examples of this behavior. The grass gets to spend more time in the rapid growth phase which can result in much more forage production.

Figure 7. Panel 1 shows an aerial view of a herd of wildebeest. The native grasslands co-evolved with vast herds of herbivores like wildebeests in Africa or bison in North America, together with their ferocious predators. Panel 2 shows a close up of the wildebeests. Panels 3 and 4 show holistic grazing of livestock using temporary paddocks to emulate nature on Sunnybrae Acres, a Canadian farm in Saskatchewan. The cattle are concentrated at one end of a paddock as they move to the next one which farmer Neil Dennis is setting up with a temporary electric fence in panel 4.

One way of achieving this with domestic animals is to emulate the predators with an inexpensive electric fence. A conventional pasture, that might formerly have been used for continuous grazing, can be subdivided in this way into 30 to 60 much smaller paddocks. At any given time all the animals are in one paddock, bounded by a temporary electric fence, where they only remain for a day or less. They don't get to return to that paddock until the grass has completely regrown. The development of automatic electric fence gate release devices called Batt-Latches has greatly reduced travel time for the farmer as the cattle have learned to anticipate when the gate will open. One can set up multiple automated gates to open in the proper sequence.

With our new understanding, we now see how herbivores like cattle, sheep, and goats can be a big part of the solution to desertification and climate change[76][77][39][40]. For more on this see the last 1/2 of my video[79] *The Magic of Soil*. This new form of grazing is very different from current grazing practices and the use of feedlots. It is not herbivores that are responsible for the damage it is how we manage them. According to the UN Food and Agriculture Organization 62% of agricultural land is used for grazing and much of that is only suitable for growing grasses that have evolved to be eaten by herbivores. By changing the way we manage herbivores to mimic nature, we have the possibility of reversing desertification, sequestering atmospheric carbon in the soil[38][39][40], and providing a much needed source of healthy food. This new form of grazing is sometimes referred to as adaptive grazing but it is more fully described as Holistic Planned Grazing. It is an important part of regenerative agriculture.

A good example of this practice is White Oak Pastures, a sixth generation, 152-year-old family farm in Bluffton, Georgia. About 20 years ago, Will Harris transitioned the farm to holistic planned grazing from a conventionally-run commodity cattle farm, that employed all of the industrial tools available, including pesticides, chemical fertilizers, hormones, and antibiotics. White Oak Pastures is now one of 47 Savory Training Hubs worldwide, where they practice, teach and study holistic management. In the process the farm went from 3 employees with medium wages to over 160 employees today with wages twice the county average and almost quadrupled the land under management[257]. It so happens that General Mills®, one of White Oak Pastures customers, became concerned about carbon footprint claims that were levelled against regenerative land management. They decided to finance an $80,000 study on a *Carbon Footprint Evaluation of Regenerative Grazing at White Oak Pastures* which was carried out by a third party environmental engineering company named Quantis[40].

To determine the carbon footprint, a life cycle assessment (LCA) was used to systematically record and analyze the impact on the environment throughout the entire life cycle of the animal. The study included emissions from the cattle like methane belches and manure emissions, as well as other farm activities, slaughter and transport, and changes in soil carbon, and plant carbon. According to Will Harris, Quantis was the same company that a competitor, Impossible Foods® (meat made from plants), used for their carbon footprint in the same time frame. The results demonstrated that for every 1 kg of fresh meat produced by White Oak Pastures, the net atmospheric CO_2 equivalent (CO_2-eq) emission was -3.5 kg. Because the value is negative this means that CO_2 was removed from the atmosphere and stored in the soil, in other words sequestered. For comparison, the result for 1 kg of plant based meat in the Impossible Burger® was +3.5 kg of CO_2-eq emission. Thus, if you eat Impossible Burger® meat you can still achieve net zero emissions by eating an equal weight of White Oak Pastures' holistically grazed meat. In their report Quantis also gives values for the net CO_2-eq emissions for conventional U.S. beef of +33 kg, for California pork of +9 kg, U.S. chicken of +6 kg, and U.S. soybean of +2 kg. The Quantis analysis assumed a methane global warming potential of 30.5 which means that a given mass of methane will trap 30.5 times as much heat in the atmosphere as the same mass of CO_2. As we will see in the next chapter the global warming effect of herbivore methane emissions are now understood to be much less than previously thought.

Savory's work, as well as research at the Noble Foundation Coffey Ranch in Oklahoma[80], found that by focusing on the grass recovery time it is possible to produce up to four times as much forage for grazing animals on the same area of land. Alberta Lamb Producers now provide a helpful fact sheet[81] on *Precision Flock Management: Forage Growth and Intensive Grazing Basics*, that takes advantage of the sigmoid shaped forage growth curve. The increased forage reduces the need to clear more forests to grow grains to feed animals in feedlots.

> **Chapter summary:**
> - It is not the numbers of animals that are responsible for desertification, it is our failure to manage plant recovery time that was achieved by nature through a balance of herbivores and predators.
> - By focusing on the grass recovery time it is possible to produce up to four times as much forage for grazing animals on the same area of land.
> - For each 1 kg of fresh meat produced by holistic grazing at White Oak Pastures, 3.5 kg. of CO_2 equivalent is removed from the atmosphere and sequestered in the soil.
> - Properly managed herbivores are an important part of the solution to desertification and climate change.

14. The new scoop on methane

But wait a minute, aren't we supposed to eat less meat? Methane emitted by herbivores is a potent greenhouse gas (GHG) but there are many other sides to this story. First, when herbivores are adaptively grazed to emulate nature there is a net reduction in GHG. The GHG emission of methane is more than compensated for by the amount of carbon sequestered in the soil[38][39][40][41][42][43].

Secondly, a new study by a global team of scientists from the Intergovernmental Panel on Climate Change (IPCC) has found that methane from herbivores is not as big a problem for global warming as previously thought[82].

What? But cattle and sheep emit methane almost constantly. However, the focus on the emissions themselves is misleading. Instead it's the warming impact of those emissions that actually matters. The methane produced by herbivores is a powerful, but short-lived, greenhouse gas. The methane is oxidized to CO_2 on a time scale of 9 years and recycled during photosynthesis to grow grass to feed the animals that breathe out methane. Similarly, we don't count the three billion tons of CO_2 humans breathe out annually because a corresponding amount CO_2 was "inhaled" from the atmosphere by the plants we consume.

This means that the methane emissions of a herd of 100 cows today are simply replacing the emissions that were first produced when that herd was established by a previous generation of farmers. There was an initial pulse of warming when the herd of additional cattle was first established, but there is no ongoing warming from that herd. There is an excellent short New Zealand video that explains the issues[83].

It's only additions to the global herd size that cause additional warming. That makes a big difference. Conversely, if the global herd size shrinks the warming decreases. In contrast, the contribution of CO_2-eq emissions from fossil fuels is not part of a short lived cycle. If we reduce fossil fuel emissions to zero, the warming will not decrease but gradually level off[84]. We actually have to remove that amount of CO_2-eq from the atmosphere to decrease the warming. One way of achieving this reduction is through regenerative agriculture which includes Savory's Holistic Planned Grazing.

It is imperative that we rapidly reduce our use of fossil fuels to prevent further warming.

Details: Until now climate science has focused on the carbon equivalence of each GHG using a metric called the 100-year Global Warming Potential, GWP100, which characterizes emissions instead of warming. There is now a new metric called GWP* which takes into account the lifetime of the particular GHG. Note: the * in GWP* is part of the name, not a footnote. According to the British Veterinary Association, the new science was well received at the COP 24 meeting in Katowice, Poland[85].

The British Veterinary Association gives an example of how the change would effect the UK[85]. Under the new updated metric, GWP*, the greenhouse gas emissions from UK agriculture fell from 46.5 Million tonnes of CO_2-eq (CO_2 equivalent) in 2016, to just 9.5 Million tonnes CO_2-eq*. (Note: CO_2-eq* indicates values calculated using GWP*, the updated Global Warming Potential.) Warming from CO_2 and N_2O across that period are the same as previously reported, but methane is recalculated as -10.6 Million tonnes CO_2-eq*. That's a negative emission value, because agricultural methane levels have fallen since the base year of 1996. By accurately measuring the impact of methane, the UK's agriculture's emissions under GWP* are just 20% of their original value.

If you are a country that has a lot of livestock farming, like New Zealand, this is good news as your GHG warming contribution has markedly decreased. If you are a climate scientist then decreasing livestock may seem like an excellent way to cool the planet by encouraging citizens to eat less meat. But, if we recognize that livestock grazed to emulate nature are capable of sequestering more CO_2-eq than they emit then the argument is very different. I will shortly discuss another reason why herbivores sequestering carbon and rebuilding the soil carbon sponge can provide a powerful lever for utilizing the water cycle to provide significant planetary cooling.

Chapter Summary:
- New science has found that methane from herbivores is not the big problem for global warming we previously thought, only increases in the global number of herbivores cause additional warming.
- Holistically grazing herbivores to emulate nature is a powerful way of sequestering atmospheric carbon and rebuilding the soil carbon sponge to help cool the planet.

15. Human planning versus nature's complexity

Everything in nature, from the microbes to the largest animals, has a role to play in maintaining our complex biosphere. We need to recognize that we are not separate from nature, but instead treasure our growing understanding of its complexity. Nature is a bank of natural capital that our very existence depends upon. Most current human management systems are rapidly depleting this natural capital. Our society is too set up to reward individuals who claim to have created a new silver bullet to solve some particular problem which almost invariably has massive unintended consequences.

Here is an example from my own province of British Columbia. For 30 years the forest industry has been using herbicides based on glyphosate, to knock down trees such as aspen and birch, which flourish in clear cuts. Glyphosate eliminates the broadleaf plants, leaving the conifers relatively unaffected. The eradication of trees like aspen and birch in regenerating forest stands is meant to make room for more commercially valuable conifer species like pine and Douglas fir[86]. Glyphosate also kills blueberry bushes and a lot of flowering species that bees depend upon and decimates large areas of forage for animals. In the process we are eliminating a lot of important diversity both above ground and below ground.

Forest Science professor at the University of B.C., Suzanne Simard, says the practice is misguided[86]. Her research has found that conifer forests don't do better when the broadleaf plants are killed off. To top it off, this leaves the more flammable conifer species vulnerable, according to Lori Daniels, a UBC forest ecologist[87]. The broadleaf plants like aspen and birch burn more slowly than the glyphosate-resistant coniferous trees, and some experts say removing them is like quite literally stoking the fires that have plagued our province. In wildfire science circles, aspen stands are often referred to as "asbestos forests"[88].

Forest herbicide contributing to wildfires was the title of a CBC News report on 21 Nov. 2019[89]. Quebec is the only province that bans the use of glyphosate in forests. The most intensive use is in New Brunswick where Prof. Rod Cumberland was fired for speaking out on glyphosate use[89].

We need to acknowledge, that in the face of the almost unlimited complexity of nature, any human solution to an isolated problem is almost certain to fail at some point. According to Allan Savory, management systems need to assume this from the outset and be constantly monitoring for the first signs of this failure so we can act proactively and replan. This is essential thinking for military planners on the ever changing battlefield. The complexity of nature, of which we are part, warrants no less. Allan Savory has spent 40 years developing such a management system. His latest book[77] on the subject written with his wife Jody Butterfield is entitled *Holistic Management: A Common Sense Revolution To Restore Our Environment*.

According to the Savory Institute website, holistic management is "a process of decision-making and planning that gives people the insights and management tools needed to understand nature, resulting in better, more informed decisions that balance key social, environmental, and financial considerations." The premise of holistic management is that nature functions in wholes. It is a holistic community with mutualistic relationships between people, plants, animals, microbes, air, water, and land. If you remove or change the behavior of anyone of the keystone species that help define the characteristics of the ecosystem, it will have a wide ranging impact, often negative, on other areas of the whole. I have attempted to provide a viewpoint of this topic from 30,000 ft, entitled *Co-creating with Nature: An Exploration of Holistic Management*[90].

As an outlier, Allan Savory's ideas have met with a great deal of resistance especially within academic circles. Not only is it a major paradigm shift in grazing practice but the holistic framework involves a larger context to consider. In particular, how we want our lives to be in the "Whole" we manage, and behaviors that will sustain that quality of life for future generations. Holistic management uses a set of testing questions to decide whether the proposed action takes you closer to or further away from your holistic context. To move forward expeditiously, the Savory Institute has set up 47 training Hubs around the world. Michigan State University is the first university accredited Hub. At the time of writing (2020) there were 124 accredited professionals, 12,546 trained land managers, and 13,275,413 hectares of land managed holistically[91].

Chapter summary:
- Reductionist management is devastating natural resources that life depends upon.
- Holistic management provides a way for humans to co-create with nature by dealing with environmental, social, and economic complexity.

PART 3
REGENERATIVE AGRICULTURE, A SUSTAINABLE FUTURE

16. Regenerative agriculture, sustainability is not enough

Earlier, I mentioned the move towards sustainable agriculture. Actually, sustainability is not sufficient because that means sustaining a badly degraded resource. A third of the planet's land is already severely degraded[92] and fertile soil is being lost at the rate of 24 billion tonnes a year, according to a 2017 United Nations-backed study that calls for a shift away from destructively intensive industrial agriculture. At the same time, agricultural soils have lost 50 to 70 percent of the carbon they once held[93]. This has contributed about a quarter of all the man-made global greenhouse gas emissions that are warming the planet. To cool the planet we not only have to end our use of fossil fuels but we also have to draw down excess atmospheric carbon.

We need to move to regenerative agriculture where we rebuild the soil biology and sequester atmospheric carbon at the same time as we grow food. Regenerative agriculture is all about mimicking nature, based on our new understanding of soil biology and nature's complexity. Charles Massy defines it as an ecological approach to farming that enables landscapes to renew themselves[94]. It's about enabling not dominating. After a transition period of several years this can eliminate the need to use fossil fuel based chemicals. Through regenerative agriculture, we have the possibility of drawing down each year as much greenhouse gas as humans are emitting[31][32][34][37][38][39][40][41][42][43][44].

The best regenerative agriculture farmers have achieved carbon sequestration rates of approximately 11 tonnes of carbon per hectare per year both with and without the use of livestock[33]. In 2010 the World Bank reported that 37.7% of the Earth's land area of 13.9 billion hectares was used for agriculture, 1.8 billion to grow crops and 3.4 billion to raise livestock. If we could achieve the best practice carbon sequestration rate of 11 tonnes per hectare per year on only 20% of agricultural land, we would be carbon neutral based on the 2019 rate of carbon emission of 11.7 billion tonnes per year, which is equivalent to 43 billion tonnes of CO_2 per year.

Using a very different type of analysis, based on ecosystem restoration or rewilding, a recent peer reviewed study by Bernardo N. B. Strassburg and 26 co-authors (Nature Oct. 2020[95]) concluded that restoration can be much more effective when it takes place in the highest priority regions. They find that restoring only 15% of converted lands in priority areas could avoid 60% of expected extinctions while sequestering 299 gigatonnes of CO_2, which is 30% of the total CO_2 increase in the atmosphere since the Industrial Revolution. They also claim that the cost effectiveness can be increased up to 13-fold when spatial allocation is optimized using their multi-criteria approach. Unfortunately, their analysis is not very explicit on the time frame for carbon sequestration which ranges from few years to centuries depending on a number of factors. All this leaves me wondering what we could achieve if we combined the best of regenerative agriculture with the best of rewilding. Of course, we still need to rapidly cut our emissions at the same time if we are to have any hope of preventing a catastrophic temperature rise.

Some of the other benefits of regenerative agriculture are greater water infiltration and storage, reduced flooding and erosion, drought-resilience, reduced use of fossil fuels (for plowing, making fertilizers and pesticides), reduced input costs for farmers, greater biodiversity both above and below ground, and more nutritious chemical-free foods. Widely recognized as the birthplace of the organic movement, the Rodale Institute in Pennsylvania has been the global leader in regenerative organic agriculture for over 70 years. Recently they have teamed up with a group of medical doctors who created the Plantrician Project. Together, they produced an informative white paper[96] entitled *The Power of the Plate: The Case for Regenerative Organic Agriculture in Improving Human Health*.

One of the Savory Institute's latest initiatives is the Land to Market program, a sourcing solution that connects conscientious buyers, brands and retailers directly to farms and ranches that are verified to be regenerating their land[97]. If we are going to reduce and reverse soil degradation, farmers, retailers, and buyers need to be assured that the methods of production are improving land health. This is the next logical step. The verification is assessed using Ecological Outcome Verification (EOV). EOV is an empirical assessment tool developed in collaboration with leading soil scientists, ecologists, agronomists, and an extensive network of regenerative land managers around the world. It is used to qualify participating farms and ranches into the program. EOV is intended as a practical and scalable soil and landscape assessment methodology that tracks outcomes in biodiversity, soil health, and ecosystem function. The latter includes the water cycle, the mineral cycle, energy flow, and community dynamics (the complex set of relationships of biology within the ecosystem). Interestingly, the more complex and diverse communities become, the more productive and stable they tend to be (more on this in Chapter 17).

EOV applies to grassland environments, including natural and planted grasslands, as well as grassland mixed with crop and/or forest areas. Farms and ranches demonstrating positively trending outcomes in land regeneration through EOV are entered into a "Verified Regenerative Supplier Roster", which participating buyers, brands, retailers and end consumers can access. EOV endorsement is bestowed as long as land health moves in a net positive direction.

Chapter summary:
- A third of the Earth's soil is already severely degraded and agricultural soils have lost 50 to 70 percent of the carbon they once held. Sustainability is the end goal after regenerating the degraded resource.
- Regenerative agriculture reverses soil degradation, improves soil fertility, and enhances the soil carbon sponge, leading to increased water infiltration and storage, drought-resilience, and reduced flooding and erosion.
- If we could achieve the best practice carbon sequestration rate of 11 tonnes per hectare per year on 20% of the global agricultural land, we would be carbon neutral based on the 2019 rate of carbon emission.

17. Six principles of regenerative agriculture

The pioneering regenerative agriculture farmer, Gabe Brown, is a frequent presenter at conferences around the world and author of the book *Dirt to Soil*[34], as previously mentioned. He is also an advocate for Allan Savory's system's thinking approach to managing resources which is referred to as holistic management[77]. Within that framework, he has distilled the operation of regenerative agriculture down to six principles. Using these principles he has increased his soil carbon from 1% in 1993 to 6.5% in 2013. During this period he increased his rain water infiltration rate from 1/2" per hour to 15" per hour. He has not used synthetic fertilizers since 2007 and his farm has become very profitable[37]. Below are Brown's six principles of regenerative agriculture and my interpretation of these principles.

Six principles of regenerative agriculture

- **Context:** Need to test any proposed action is within our ecological, financial, community, and spiritual context. Holistic context is how we want are lives to be and the environment and behaviors that will sustain that quality of life for future generations. Context lies beyond any immediate goal or problem.
- **Limit disturbance:** Limit mechanical and chemical disturbance of soil. Tilling causes a lot more soil carbon to be converted (oxidized) to CO_2 and destroys soil structure built by the microbes, leading to soil erosion and reduced water infiltration.
- **No bare soil:** One role of plants is to cover soil whether dead or alive. Its litter acts as armour to insulate the soil surfaces against weather, preventing drying out and erosion. Litter provides a bed and breakfast for fungi and other organisms including earthworms.
- **Build diversity:** Nature never has monocultures. Dr Ademir Caligari, a Brazilian scientist and a leading expert on cover crops, inspired Brown to sow 15-25 simultaneously. Above ground diversity provides a greater diversity of root exudates to support a greater diversity of soil microbes. Nature achieves nutrient cycling with this biology. Synergies compound once 7 or 8 plant species are grown together in a "cocktail." (Professor David Tilman and collegues have established many benefits to biodiversity[98]. Their award winning research is illustrated in Figure 8.)
- **Keep living roots in the ground:** Maximizes photosynthesis. Brown extends his 100-day growing season by sowing species from all four groups – cool season grasses and broad-leaves, warm season grasses and broad leaves. Living plants produce exudates to feed soil life which renew soil aggregates.
- **Integrate animals:** Nature does not function without animals. Livestock are nature's mobile biodigesters and biofertilizers. The grazing of plants stimulates plants to pump more carbon exudates into the soil. This drives nutrient cycling by feeding soil microbes. Brown uses holistic planned grazing with very high stock densities[34], rotated quickly through many paddocks of multispecies cover crops as a tool to gain a big improvement in soil health.

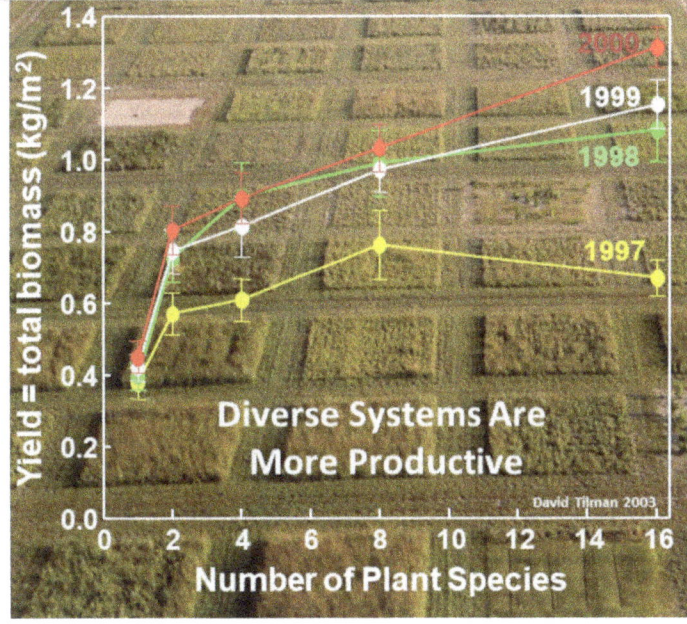

Figure 8. This figure summarizes the many benefits of biodiversity that have been demonstrated in experiments carried out at the University of Minnesota Cedar Creek Ecosystem Science Reserve by Prof. David Tilman and colleagues. (Permission granted to use image and figure by David Tilman.)

18. The global thermostat and how to cool the planet quickly

Wow, this topic deserves to be the subject of a whole book. Let me see if I can do it some justice to the topic in 4 pages. It is often said that water vapour is a much more important greenhouse gas than CO_2, so why all the focus on CO_2? In 2010, a group of NASA scientists, Andrew Lacis and colleagues, wrote a paper entitled *Atmospheric CO_2: The Principle Control Knob Governing Earth's Temperature*[99][100]. They used their best climate model to investigate what would happen if they set the levels of the non-condensing greenhouse gases (GHG) to zero. These include CO_2, methane, nitrous oxide, ozone, and chlorofluorocarbons. These GHGs do not condense and precipitate from the atmosphere at current climate temperatures, whereas water vapour can and does.

What Lacis and his colleagues found was that with only water vapour included in their model, the terrestrial greenhouse effect would collapse. After 50 years, the model predicts a global temperature of -21 °C, a decrease by 34.8 °C, plunging the global climate into an icebound Earth state. They concluded that the non-condensing greenhouse gases, which account for 25% of the total terrestrial greenhouse effect (CO_2 alone contributes 20%), serve as the principal control knob that governs the Earth's temperature. It is this control knob which sustains the current levels of atmospheric water vapour and clouds that account for the remaining 75% of the greenhouse effect.

The conclusion is that we need to reduce but not eliminate the non-condensing GHGs like CO_2 to turn the temperature down. Sequestering atmospheric carbon in the soil is a natural part of regenerative agriculture which offers so many other benefits like rebuilding the soil carbon sponge. Efforts are also under way to develop technology to remove CO_2 directly from the atmosphere[110]. Unfortunately, reducing CO_2 level in the atmosphere will be slower than you might expect because the oceans have been absorbing about 40% of the human caused CO_2 emissions to date. This has led to ocean acidification. Once we start withdrawing CO_2 from the atmosphere the oceans will release some of their excess carbon back into the atmosphere in a rebalancing. The good news is that regenerative agricultural practices provide other opportunities to cool the planet while we work at reducing greenhouse gases. Let's explore some of these.

Walter Jehne, an Australian microbiologist and climate scientist, has drawn attention to the importance of rebuilding the soil carbon sponge[101][102][103][104] to provide a powerful lever for influencing the water cycle, which, as we have learned, controls 75% of the global heat dynamics. Jehne is an interesting outlier in the climate change discussions who emphasizes the important role of biology in climate dynamics. Jehne argues that an enhanced soil carbon sponge will provide opportunities to cool the planet through changes to the water cycle.

Jehne starts from estimates of the imbalance between the sunlight energy absorbed at the top of the earth's upper atmosphere and the thermal (heat) energy the earth radiates back into space from the top of the atmosphere. When the two are equal there is no net energy heating or cooling the planet. At this time they are not equal because the greenhouse blanket effect is increasing, so more of the thermal energy is being trapped causing the planet to warm. Currently, the radiation flowing in from the sun is greater than the radiation leaving the earth, so the earth is warming. This difference is called radiative forcing because it is forcing a change in the earth's temperature. Estimates of this forcing range from about 1 W to 3 W for each square metre of the earth's surface area averaged over all latitudes and day and night. Jehne uses the 3 W per square metre value which represents less than 1% of the solar radiation incident on our planet. Even if the GHG stabilized at a new higher level than in the past, the planet's temperature would continue to rise ever more slowly for about another 100 years[105]. So what can we do to cool the planet?

Jehne identifies many ways that we can produce a cooling effect to compensate for the 3 W per square metre imbalance in a timely manner through changes to the water cycle. Here is how it works. Recall that soil microbes build soil structure (the soil carbon sponge) with fungal strands plus the glues they make from the carbon compounds they receive as exudates from the plant roots. These soil structures are full of voids for storing water. Jehne calls these voids "cathedrals" and in a good soil with 5% organic matter (2% - 3% organic carbon content) the voids amount to about 50% by volume whereas in excellent soil with 10 to 18% organic matter the voids can reach 70%[106].

Soils with these voids are what we call the soil carbon sponge. They can infiltrate and store a great deal of water. This sponge buffers the effects of extreme weather peaks by minimizing flood peaks, erosion, storm damage, and extending the longevity of green growth through periods of drought. More plants mean more photosynthesis, more exudates, more microbes, and improved nutrient cycling. Those extra plants also mean more evapotranspiration of water and enhanced cooling through the latent heat of vaporization. At 25°C, 590 calories of heat are removed from the environment when each gram of water evaporates or transpires from a plant to become water vapour. The more trees and plants that are able to grow and transpire, the greater this natural air conditioning will be[107].

In one experiment using an infrared temperature gun, the temperature of a living plant leaf was compared to that of an adjacent piece of green paper exposed to the same sunlight. The leaf had a temperature of 15°C while the green paper measured 32°C. The air temperature was 19.4°C. Unlike the paper, the leaf is transpiring water which is the dominant cooling mechanism. In addition, photosynthesis stores about 10% of the incident sunlight in the carbon bonds of the carbohydrates they produce[120] which also cools the leaf. In another experiment, when I measured the temperature of the soil surface under a full canopy of green plants in our garden, it registered 18°C, the same as the air temperature. When I measured the temperature at the surface of a patch of bare soil it was 44°C. In much more arid places where there is very little evaporative cooling of the bare ground, the surface temperature can hit 60°C. According to Jehne[102][103], the enhanced radiation from the hot bare ground heats the air and can produce a high temperature heat dome that inhibits moist low pressure air from entering the region causing greater aridification. He cites as examples of this phenomena, California's Central Valley and the interior of Australia.

Currently, some 25% of the incident solar radiation is naturally transferred away from the surface of the earth through these evapotranspiration latent heat fluxes[108]. If we are able to increase the quantity or longevity of green plant transpiration by only 4%, we can offset the abnormal warming that we have induced today of 3 Watts per square metre. According to Jehne, there is no danger that we would over cool the planet because if we cooled it too much, photosynthesis and transpiration will slow down and self regulate.

To put this in perspective, approximately 40% of the earth's land area is already desertified or desertifying. I am persuaded that the Savory Institute has provided good evidence that we can re-green much of the desertifying landscape using holistic grazing. This is nicely illustrated in Figure 9.

Figure 9. In the 1970s, Allan Savory helped the Kroon family in the Karoo Desert of South West Africa (Namibia) to turn the desert on the right back to the grassland on the left[86a]. © Savory Institute, reprinted with permission

In conventional agriculture the practice of leaving a field fallow (bare with nothing growing such as a cover crop) for a season is very common especially in more arid regions. The more plants and trees we have covering the surface, the more cooling we can achieve. This cooling will be very important to global food security. During the 2003 extreme heat event in Western Europe, the July-August temperature was 3.6°C above normal. Corn (maize) yields dropped by 36% in Italy and by 30% in France[109]. France also saw a 25% reduction in fruit harvests and a 21% reduction in wheat.

Highly reflecting clouds provide another cooling mechanism. They reflect approximately 22% of the incident sunlight back into space[108]. Increasing cloudiness by 4% is another way to offset the abnormal warming we have induced today. Of course, with the increased plant transpiration we have added more moisture in the air capable of producing more clouds. According to Jehne, an important factor in nucleating cloud droplets are microbes that are carried up into the atmosphere from the leaf stomata of certain trees and plants.

It is important to note that the two cooling processes described above are a direct result of regenerative agriculture's ability to restore the soil carbon sponge leading to increased soil moisture. The same benefits are not expected from current technological efforts to remove CO_2 directly from the atmosphere[110]. Even if they can be made to work at scale, the end result is either to turn CO_2 into rock underground, or convert it to carbon-neutral synthetic diesel and petrol. Neither will build the soil carbon sponge to provide the leverage to cool the planet by enhancing the water cycle.

It looks like biology plays a significant role in controlling the water cycle[111][112][113][114][115][116]. It is not simply atmospheric physics. To buy us time the carbon sponge is essential to rehydrate the land allowing for much more and longer duration of plant growth, more cooling transpiration, and more tree based aerosols (including microbes) capable of nucleating clouds, precipitating rainfall and providing the biotic pump[115][114] necessary to draw moisture laden ocean air inland. To see whether bacteria can act as cloud condensation nuclei a team of scientists collected snow and ice samples from many sites around the world: France, Montana, the Yukon, even Antarctica. Of the 19 fresh snowfalls they analyzed, they found that biological ice nucleators were everywhere[117].

Microbes are amazing and the sooner we learn to cooperate with them the better are our chances of survival. Recent research found microbes that have been wafted high up into the atmosphere. David J. Smith, a research scientist at NASA Ames, identified 2100 different types of living microbes at an elevation of 9000 ft[118]. More than 27% of the bacterial samples and more than 47% of the fungal samples were still alive at this altitude.

Jehne provides a table of actions and estimates outlining how we can readily and practically draw down 20 billion metric tonnes of carbon per year which is about twice the current annual increase of carbon in the atmosphere[119]. To convert metric tonnes of carbon to metric tonnes of CO_2 multiply by 3.67.

Chapter summary:
- There are two types of greenhouse gases (GHGs), (a) the non-condensing gases that include the well known CO_2, methane, and nitrous oxide, and (b) the one condensing GHG, water vapour.
- The non-condensing GHGs, dominated by CO_2, are the principal control knob governing Earth's temperature. It is hugely important that we move quickly to stop adding fossil-fuel based GHGs.
- These non-condensing GHGs sustain the current levels of atmospheric water vapour and clouds that account for 75% of the greenhouse effect.
- It is important to draw down the excess atmospheric carbon. In the process the oceans will release some of their excess carbon into the atmosphere which will be good for the oceans but slow our efforts to reduce the level of greenhouse gases.
- Sequestering carbon in the soil using regenerative agriculture has the added advantage that it rebuilds the soil carbon sponge. This can provide an important lever for influencing the powerful water cycle to rapidly cool the planet in a variety of ways.
- Microbes play a key role in building the soil carbon sponge and they also appear to be important in seeding rain clouds.

19. The ABCD of agriculture

Walter Jehne provides a simple way of grasping the benefits of regenerative agriculture compared to conventional agriculture[101][102][103]. He calls it the 'ABC' of agriculture. His presentations are always done by writing and sketching on large easel pads or whiteboards. I have created Figure 10 to illustrate some of his key points using language and terms already discussed in this book.

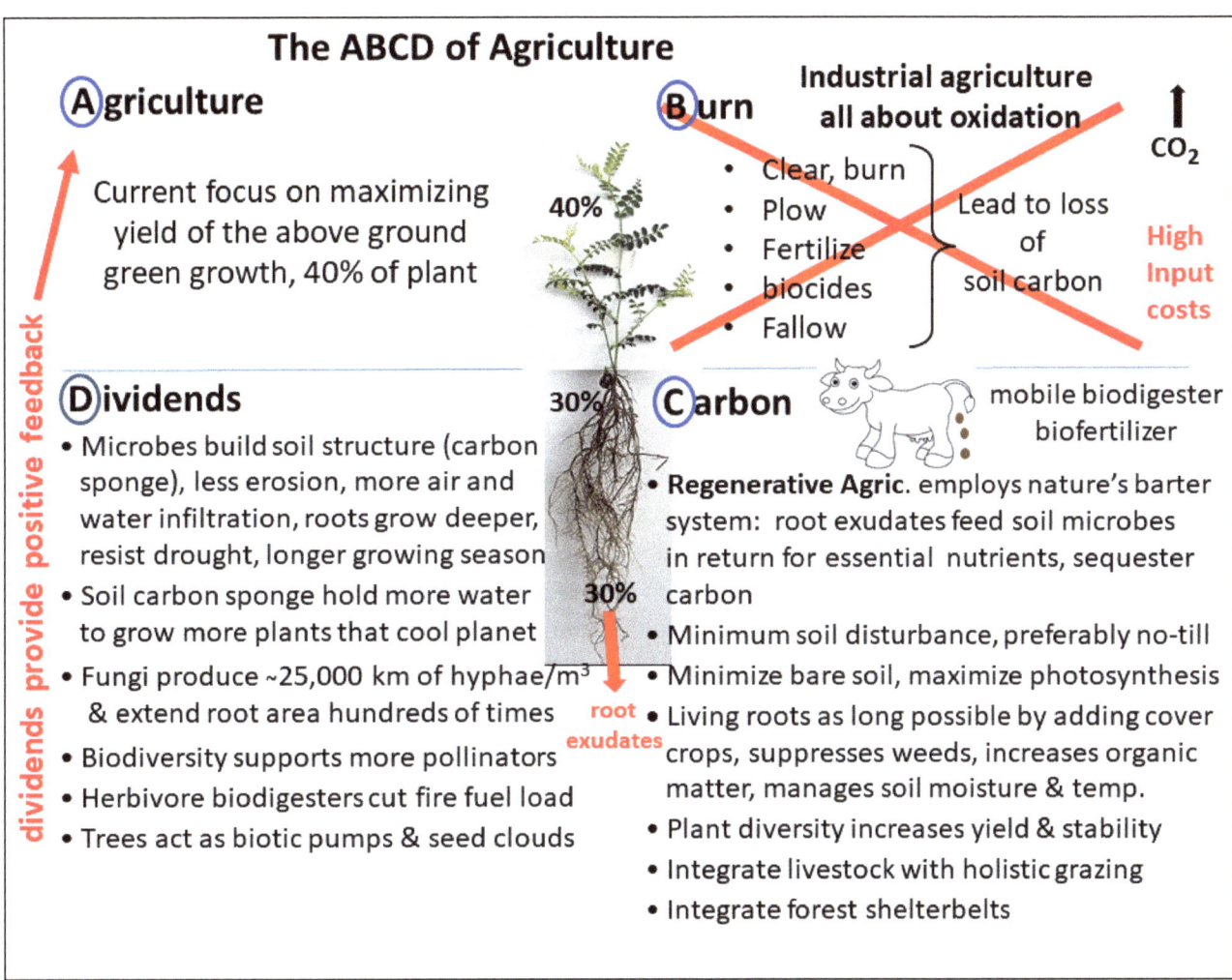

Figure 10. On the left 'A' stands for Agriculture. On the right 'B' stands for Burn representing the unsustainable oxidative practices of industrial agriculture leading to increased atmospheric carbon dioxide emissions and the rise in farm bankruptcy because of high input costs. The 'C' stands for the all important soil Carbon. The bottom right panel represents the practices of regenerative agriculture, practices that mimic nature and sequester carbon with the help of soil microbes. In this example, approximately 30% of the sugars created by photosynthesis are released by the roots to power the soil microbes. Herbivores like cattle are nature's mobile biodigesters and biofertilizers. The 'D' stands for Dividends provided by regenerative agriculture that increase soil health and stability and provide positive feedback to enhance above ground plant yield and nutrient content.
(figure by Phil Gregory, inspired by a 2016 talk by Walter Jehne[102])

20. One man's journey through agriculture

In the course of researching this book, I interviewed Dr. Kent Mullinix, the Director of the Institute for Sustainable Food Systems at Kwantlen Polytechnic University in British Columbia. His career journey is interesting because it provides a more intimate close-up picture of developments in agriculture from the 1970s to the present.

Kent Mullinix grew up in a military family, disliked school immensely and experienced high school in Montgomery Alabama in the late 60s and early 70s, soon after forced desegregation. With no prior farming experience, Kent enrolled in the University of Missouri and finally found his passion in horticulture, completing a B.S. (1976) in agriculture, specializing in fruit and vegetable production. He graduated with an award from the department as an outstanding undergraduate student with a passion for agriculture, not just plants. He continued his education at Missouri for an M.S. and a PhD in agricultural education, with a few years in between as a research specialist at the Universities of Minnesota and Kentucky. Kwantlen is now the fifth university that he has worked for in agriculture, as a researcher and as a professor. For his whole career, he has lived and worked in agricultural communities. For Kent it has been an incredibly intriguing and rewarding career, "I eat, drink, and sleep agriculture."

Kent's training followed the dominant industrial agriculture model of monocultures, fossil-fuel based fertilizers and chemical pesticides. The concept of sustainability first emerged in agriculture when he was finishing his undergraduate degree. It was called low input sustainable agriculture or LISA and Missouri was a focal point for this. Kent explained that LISA was championed by some agricultural academics but they lost out to the green revolution corporate interests.

During this time he became aware of the plight of family farms in Missouri being driven into debt by industrial agricultural practices and the consequential collapse of farming communities. During graduate school, Kent and his wife lived on a family farm that went out of business during the market adjustment of the 80s. He described them as the most dedicated farmers you could ever imagine, a 230 acre mixed cow-calf, hog, corn, wheat, hay operation run meticulously and beautifully and they went out of business because they couldn't compete.

"Anyway it was the social disruption, the disruption of the agriculture that I fell in love with as a twenty year old, it was vanishing in front of my eyes, " said Kent. He also witnessed the integration of industry and politics. "The agricultural industry started advancing their global industrial vision by becoming very politically active and aligning themselves with the republican party: the downsizing of government, the elimination of regulation and stewardship, and the abandonment of fair wages." According to Kent, the same thing has happened in Canada. "With the whole sale embracing of industrialization and deregulation and the lack of care for the environment, community, and workers, everything became a commodity, everything was about making money, nothing was about our communities, people, the health of the earth and environmental stewardship."

It wasn't just in agriculture. Kent remembers seeing Walmart come into a small Missouri town and devastate it. The hardware store boarded up, the grocery store boarded up, the local clothing store boarded up. Along with that and the devaluation of family based farming in these communities, he saw the tractor dealers, the feed suppliers, the implement dealers, and the car dealers, all go out of business.

After witnessing all that, he got a position in Eastern Washington where the tree fruit industry was still dominated by small family farms. "You could make a living with 10 acres of Golden Delicious that were just for the apple sauce market," said Kent.

The bulk of his career was spent in the tree fruit industry of Eastern Washington which is more than 30 times the size of the fruit industry in British Columbia. There he set up a two year educational program to train horticulturalists for the fruit industry at Wenatchee Valley College, that ultimately became the first two years of a Washington State University undergraduate degree. It was cited in the U.S. as the best technical education program of any discipline in the nation. The program was focused on preparing farmers and professional horticulturists/tree fruit specialists but he was bucking a trend. He explained that most agriculture programs in universities today are not concerned with farmers. They actually don't think we need farmers but rather managers and business people for the industrial food system because after all family farms can no longer make a living.

I asked Kent why that was? "I became really concerned, within the context of sustainability, of what we call the technological treadmill. This is really what is driving much of the adverse economy of family based agriculture. It started before the green revolution technologies, but it really ramped up with the introduction of farming systems based on hybrid seeds, fossil-fuel based fertilizers and fossil-fuel based pesticides and irrigation. When farmers invested in these technologies, which were controlled by a handful of corporations, their cost of production started going up, up, up. And their production yields started going up, up, up too, because that's what this as about. Simultaneously, it resulted in overproduction and the depression of value and the margin. So it put farmers on this treadmill. They had to keep adopting technology after technology just to continue to produce enough to chase this minuscule margin to stay in business. It's folly because it's an economic trajectory of only one winner and that's what we are seeing in family farming. Thinner margins meant farms needed to become much bigger. Farmers end up predating on their neighbours just to economically survive." This is the story behind Darrin Qualman's dramatic graph shown in Figure 1 of Chapter 3.

Kent continued, "As an agricultural scientist and a pomologist [fruit and nut tree specialist] I am a person training the agricultural leadership of this sector. I was becoming increasingly disturbed that the industry was becoming entirely dependent on the profligate use of pesticides that we all knew were toxic to humans and off target species, a plethora of off target species. Simultaneously, I'm watching the industry get ticked off and politically active when environmental activist groups called them on it, saying that apples are covered with pesticides and we're feeding them to our children."

Instead of responding to consumers concerns and responding to environmental scientists and ecologists about biodiversity destruction, they fought it. Kent found this very disturbing. He became really focused on how to do things differently, and began studying Integrated Pest Management (IPM). Between 1945 and 2000, despite the more than 10-fold increase in insecticide use in the U.S., total crop losses from insect damage nearly doubled[187] from 7% to 13%. Kent realized that our use of pesticides creates the vast majority of pest problems.

One prominent scientist that influenced Kent's thinking was W. J. Lewis of the Insect Biology and Population Management Research Laboratory at the USDA Agricultural Research Services in Tifton, GA. In the abstract of a 1997 paper on *A total system approach to sustainable pest management*, Lewis and colleagues wrote[188], "Pest management strategies have long been dominated by quests for "silver bullet" products to control pest outbreaks. However, managing undesired variables in ecosystems is similar to that for other systems, including the human body and social orders. Experience in these fields substantiates the fact that therapeutic interventions into any system [in agriculture, killing pest organisms with toxic chemicals] are effective only for short term relief because these externalities are soon "neutralized" by countermoves within the system. Long term resolutions can be achieved only by restructuring and managing these systems in ways that maximize the array of "built-in" preventive strengths, with therapeutic tactics serving strictly as backups to these natural regulators. To date, we have failed to incorporate this basic principle into the mainstream of pest management science and continue to regress into a foot race with nature."

As part of his research Kent took a sabbatical at the University of British Columbia Faculty of Agriculture (now Faculty of Land and Food Systems) to study Integrated Pest Mangement with Murray Isman, an entomologist. The research project that Kent undertook for his 2004 UBC PhD, was to document the arthropod population dynamics, e.g., insect pests and "beneficials" in apple orchards cultivated with and without the standard regime of pesticides. He compared the standard pesticide orchard management approach to a biological approach. It was a replicated orchard scale study (acre blocks) carried out in Washington State. He followed the dynamics of the arthropod populations and looked at the damage they did and the economics of the two approaches. It was a four year experiment collecting data on 15 different arthropod species. One result was that every pest in the orchard, except one, became a non pest with the elimination of pesticides. The non pests did not cause economic damage. They only became pests when their enemies in the natural ecosystem were eliminated using a pesticide regime.

The one pest that remained problematic was the codling moth (an exotic with few natural enemies) which is the pest that is responsible for the classic worm in the apple. They combined three biological approaches that had been developed to manage this moth. The first is called mating disruption which uses a pheromone to mimic the female pheromone that attracts males for mating. When emitters are hung in the orchard at a certain concentration the males can't find the females. They also used a virus called granulovirus which is specific only to this species and causes a rather high mortality rate. The third method, referred to as sanitation, is simply looking for apples that have become infested and removing them from the orchard in a timely fashion before the pupa can drop to the orchard floor for over wintering.

Kent's study successfully demonstrated that the biological approach was as effective as pesticides and more profitable. "From a production standpoint the yields were equal, the pest damage was equal, so eliminating pesticides did not exacerbate damage, and the economics were better." This is reported in the journals and his thesis[189][190]. Well, his results were ignored. He believes they were ignored because it is not in the mindset of what the industrialists want to do. According to Kent, university departments of agriculture and pomology are full of entomologists that are not sufficiently dedicated to changing the system. They are largely working to advance industrialization. This is reflected in the unfortunate dominant interpretation[191] of IPM as 'Integrated Pesticide Management'. There is very little personal advancement in eschewing pesticides and embracing an ecological agroecosystem approach as promoted by the UN Food and Agricultural Organization. In Kent's view, "Ultimately, any of us that study ecology know that humanity is destroying mother earth and we are at the 11.5 hour."

The upshot of his UBC PhD work and focus on sustainability was that he was increasingly ostracized in Washington and eventually gave up his tenured professorship at Washington State University to move to Canada for a two year position at Kwantlen College. An opportunity arose in 2008 when Kwantlen College became Kwantlen Polytechnic University and Kent was appointed the Director of a new Institute for Sustainable Food Systems, an applied research and extension unit. Soon thereafter, he led the development and implementation of KPU's Sustainable Agriculture Bachelor of Applied Science degree; that is the only one of its kind in North America. This degree program even has one of the first courses in sustainable economics. Kent invited Missouri University professor emeritus of agricultural economics, John E. Ikerd, to give the first course of lectures. When they first met many years prior, Ikerd confided that early in his career, as an extension agriculture economist, he advised farmers they couldn't survive without adopting the industrial approach. Years later he had a change of heart when he deduced that many of the farmers that had adopted the industrial system had gone bust. In other cases the farmer is reduced to being a corporate hired-hand by signing a comprehensive production contract.

Ikerd's story reminds me of the interview I did with fifth generation organic cattle farmer Harold Steves, who is widely considered the father of the British Columbia Agricultural Land Reserve (ALR). Harold studied genetics at UBC around 1960, where he was exposed to two professors with completely opposite views on the future of agriculture. One warned that farmers that didn't convert to chemical agriculture wouldn't last. Harold chose to follow the advice of Professor McKenzie who advocated the organic point of view that was bucking the trend. All of Harold's friends that went into chemical agriculture left farming. Harold is a big proponent of regenerative organic agriculture.

Kent is passionate about a small farm based, deeply sustainable regional food system, and believes we need to move from 1.5% of the population involved in farming to 20% as we move off fossil fuels. According to Kent 1.5% is not even enough for biological replacement and requires us to import temporary farm workers from other countries.

I asked Kent about funding for agricultural research. His answer, "Further industrialization and globalization is the focus of virtually every agricultural university and most federal funding streams. That is what it is about. Funding streams that are genuinely focused on advancing deeper sustainability are minuscule." Because his institute is not focused on the mainstream industrial agriculture paradigm they have not been successful garnering research funds from the provincial agriculture research branch which is called the Investment Agriculture Foundation (IAF). Funding awards generally require a 25% industry match. As described on their website, IAF is an industry led, not for profit that delivers government funded programs to the agriculture and agri-food sector of British Columbia. There is little opportunity to get any of that funding for something like small scale, regenerative agriculture that doesn't fit into the dominant industrial agriculture model and that does not have a lot of resources to devote to research. In my view, this type of government funding provides a powerful positive feedback lever for corporations to maintain the industrial model of agriculture and control the flow of tax payer money to their benefit. In spite of these formidable hurdles, the Kwantlen Institute for Sustainable Food Systems has been very successful with funding from private foundations which are deeply concerned about the environment and a just sustainable equitable society.

Recently, the University of Illinois launched a Regenerative Agriculture Initiative[192] funded by Fresh Taste[193], a group of Chicago-based foundations and city officials interested in changing how food is produced, processed, and consumed in the Chicago region. Fresh Taste now has more than ten members and have awarded over $32 million for food system work from 2007 to 2017. So perhaps we are seeing an encouraging trend of foundations stepping up to encourage the needed paradigm shift to regenerative agriculture.

Chapter summary:
- The short-term business interests of the global food industry are not aligned with the true needs of farmers, consumers, society, and the environment.
- To date, we have failed to incorporate basic ecological principles into mainstream pest management and continue the futile effort to dominate nature with an arsenal of toxic chemicals.
- Our use of pesticides creates the vast majority of pest problems.
- Our unsustainable global-industrial food system didn't just happen. It is the result of collaboration between industry, government, and university.
- An alternate resilient regenerative food system won't just happen either. It will require a purposeful collaboration between farmers and other sectors deeply concerned about the environment and a just sustainable equitable society.

21. The true cost of food

The picture of agriculture that emerged in my six year investigation is like a heavily weighted coin. One side is the dominant industrial model which in Canada has efficiently transferred approximately 97% of farming revenue (see Chapter 3) to the pockets of multinational corporations, largely decimated family farms and rural communities, poisoned the environment, enhanced greenhouse gas emissions, and created cheap unhealthy food. It is an unsustainable approach to farming that pits human technology against the complexity of nature and measures success only in short term economic returns. Nevertheless these corporations have managed to achieve effective control of agri-food public policy and educational messaging to extend the lifetime of a paradigm that threatens our continued existence.

On the other side of the coin is a deeply sustainable ecological approach to farming, called regenerative agriculture, that enables landscapes to renew themselves even if they have been badly degraded. It is based on a revolution in our understanding of soil biology and nature's complexity that has emerged in the last 3 decades. It is an approach that replaces expensive chemical inputs with the biological systems nature has evolved over hundreds of millions of years. This new knowledge enables us to co-create with nature. By that I mean interact with nature in a way the rebuilds or regenerates the natural capital going forward in time. This enables nature to do a lot of the work for us when it comes to producing our food and building soil. It is nothing short of a complete paradigm shift that has the potential to deal with many of the existential threats that humanity faces including the imminent collapse of agriculture, the collapse of biodiversity, climate change, and an explosion of chronic disease epidemics. However, paradigm shifts are extremely difficult to bring about because existing institutions have built in reward systems that reinforce the existing paradigm.

To help create a more sustainable, more nutritious, and more just food system, many organizations[194] are attempting to account for the true cost of our food. This means including the externalized costs associated with a poisoned environment, enhanced greenhouse gas emissions, and health issues created by cheap unhealthy food and working conditions. In addition, food prices don't reflect the cost of essential public services paid by tax payers for things like highways, legal systems, and a variety of government farm subsidies. To estimate the hidden costs of food it is necessary to look at a wide range of social, human, and environmental factors. This approach is called True Cost Accounting.

One of these organizations is the UK Sustainable Food Trust which organized an important conference on *The True Cost of American Foods* in San Francisco in April, 2016. According to agricultural economist Professor Emeritus John E. Ikerd[195], who spoke at the conference, "Economic estimates may be essential in bringing public attention to the importance of economic externalities. However, public policy initiatives and political movements must rely primarily on social and ethical values. Relying solely on market incentives allows ethical decisions to be decided by one-dollar-one-vote, rather than one-person-one vote. Policies should not allow what is ethically unacceptable as with slavery in the past. Today, animal welfare ought to be protected."

21. The true cost of food

In November 2017, the UK Sustainable Food Trust produced a report[196] on the True Cost of UK Food in which they estimated the hidden cost was equal to the retail price. The report was revised and updated in 2019, which only slightly adjusted the hidden cost to 96.8p for every 100p spent.

The true cost varies according to how the food is produced and how well or poorly it contributes to a healthy diet. Does farm size affect the true cost of food? For John Ikerd[197], the short answer is yes!

"Industrial agriculture degrades and destroys the relationship between the farmer and the land. Specialization, simplification, standardization, and mechanization allow each farmer to cover more land, supervise more workers, and handle more dollars. --- But, as the attention of each farmer is spread over more land, more labourers, and more capital, each acre of land, each worker, and each dollar receives less personal attention. The relationship of the farmer with the land, and with the people of the land, is weakened. If the large farmer no longer knows the land, no longer cares about it, forgets how to care for it, doesn't have time to care for it, or can't afford to care about, how well will the land be used? How can it remain productive? How can a large farm be sustainable?"

The increase in size of farms in the US is a result of a quest for economic efficiency, which inevitably conflicts with ecological and social integrity. As farms have grown larger, the external economic costs of farming have risen, suggesting a relationship between farm size and economic externalities. Ikerd concludes that today's large farms are the right size for economic efficiency but they are too large for ecological, social, and economic sustainability[197]. He provides compelling arguments for small farms intensively managed for sustainability. For further details of his arguments see his keynote addresses to the MOSES Organic Farming Conference in 1999[197] and 2020[198].

As we discussed in Chapter 16, sustainability may not be enough because that can mean sustaining a badly degraded resource. Agricultural soils have lost 50 to 70 percent of the carbon they once held[93]. Fortunately, regenerative agriculture provides a way of reversing soil degradation and returning to nature's natural sustainable cycles.

The Sustainable Food Trust report also indicates that in general, true cost accounting has not yet been applied to many social, cultural and ethical issues associated with food systems. Some of the social issues include the decimation of family farms and the rural communities that depend on them. With the move away from mixed farming to specialization in either crop or livestock production, there are now few full-time jobs available and most of the demand is for seasonal workers, most of whom are now migrants. Many of these seasonal labourers are often paid less than national workers. The report indicates that worker exploitation has become common in agriculture due to the pressure to produce and sell cheap food and no one has attempted to calculate the cost to society from the declining agricultural labour force and the major changes in the type of work available over recent decades. According to Ikerd, these non-economic external costs of large farms may matter even more than the economic externalities.

I will close this chapter with a final quote from John Ikerd[199], "Sustainable farms must meet the basic food needs of all the present generation, without diminishing opportunities for the future. Today's large farms obviously are doing neither. The percentage of people in the US classified as "food insecure" today is about three times larger than during the 1960s. Today's industrial food system is linked to an epidemic of obesity and other diet-related health problems. In addition, today's dominant farming systems are degrading the health of soils and mining the productivity of the land and demeaning the agricultural workforce – resources essential for the future of agriculture."

Chapter summary:
- Today's agriculture is like a weighted coin. On one side is the dominant, entrenched but unsustainable industrial model which threatens our continued existence. On the other, a sustainable ecological approach working with the biological systems nature evolved called regenerative agriculture.
- True Cost of Food initiatives provide a way to assess the hidden costs of food, highlighting the externalized cost arising from a poisoned environment, enhanced greenhouse gas emissions, unhealthy food, and poor working conditions.
- The UK Sustainable Food Trust estimates that the hidden costs are approximately equal to the market price. They also recognize that their estimates do not currently include many social and ethical issues which may matter even more than the current economic externalized costs.
- The industrial approach leads to larger farms that are the right size for economic efficiency but are too large for ecological, social, and economic sustainability.

22. Bountiful small regenerative farms

One notable example of a successful, intensively managed small farm using regenerative agricultural practices is Singing Frogs Farm[200][201][202] in Northern California, run by Elizabeth and Paul Kaiser. It is an eight acre farm which grows food on just under three acres. Since starting in 2007 they transitioned from tractor powered tillage to raised beds and a rototiller, to no-till/no dig using a broadfork as their next transition tool. During this time their soil organic matter went from 2.4% to between 8% and 11%. Since 2013, they have not used a broadfork on most of their farm because it is no longer necessary now that the soil organic matter is so high. They grow year round and have 3 to 8 sequential economic crops in any one bed each year in spite of the fact that their frost free growing period is only 110 days. 98% of their employee's labour is transplanting and harvesting. They really appreciate the important role that soil biology plays and follow the principles of regenerative agriculture (see Chapter 17). They use no sprays for pests, weed control or other issues and apply only compost and organic soil amendments. According to Paul, no-till is vastly different from the past 10,000 years of farming in that one has to really commit to understanding soil science, soil biology, and ecology.

Their farm also benefits from a variety of domesticated animals including a flock of free-range chickens for eggs and pest control, a flock of sheep (rotational lawn mowers), a milking goat, a farm dog, and barn cats. They also harvest a wide variety of fruit trees and blueberry bushes, and regardless of the season they have a bounty of fresh produce[228]. Singing Frogs Farm uses direct marketing through CSA shares (Consumer Supported Agriculture), local farmers markets and restaurants. As of April 2020 they had 140 CSA shares[228], with a classic box costing $26 USD per week and a family box costing $33.50.

Their small farm produces an immense amount of food with a gross revenue of $100,000 USD per acre per year and a net revenue of $85,000 per acre which is enough to support 5 to 6 full time equivalents working 40 to 50 hours a week. In recent correspondence Elizabeth Kaiser (17 Dec. 2020) reported that the farm is on track for a gross revenue of $145,000 USD per acre in 2020. As there are typically 16 people involved in the farm this allows for vacation time as well. In their presentations the Kaisers provide comparison gross revenue statistics for California vegetable farms. The average for all California vegetable farms is $1,900 per acre per year. This rises to $3,700 for organic vegetable farms and to $11,000 for small, diversified, direct market organic vegetable farms and Sonoma County vineyards. For economic viability, the revenue of small, diversified organic vegetable farms in California is considered to be $14,000 per acre per year or better. This means a lot of people aren't making it but are enjoying homesteading or gardening. 95% of Singing Frogs' produce stays within about a 20 miles radius of their farm. They see a bright food resilient future with many small farms like this located close to population centres.

The gross revenue per acre of Singing Frogs Farm is a remarkable 53 times higher than the average of all California vegetable farms. It would be nice to know what fraction results from an increased yield per acre arising from Singing Frogs Farm's intensive regenerative agriculture management practices. The other factors determining the gross revenue per acre are product price to the end consumer and market share received by the farmer. In the panel below I give my very rough back-of-the-envelope calculation aimed at estimating the ratio of Singing Frogs farm's yield per acre to the yield per acre of the average California vegetable farm.

Details of the back-of-the-envelope calculation

Below is the equation we want to solve for the **YieldPerAcreRatio**.

GrossRevenuePerAcreRatio = YieldPerAcreRatio × ConsumerPriceRatio × FarmerMarketShareRatio

- **GrossRevenuePerAcreRatio** = 53 times from above.
- **ConsumerPriceRatio** = price ratio experienced by the end consumer of the produce. For this very rough calculation I assume this ratio = 1.

The end user for the conventional farm produce is assumed to purchase the produce mainly from a grocery store at the local retail price or from a food service provider like a restaurant. Singing Frogs Farm mostly markets directly to the end consumer and according to the Kaisers they price their produce at the local retail price. A portion of their produce is purchased by restaurants.

- **FarmerMarketShareRatio** = 8.

The basis for this rough estimate is a 2018 Washington Post article[229] on what fraction of every dollar Americans spend on food is captured by the farmer. Keep in mind we are only considering gross revenue. For this calculation we do not need to consider the farm's expenses to produce the products. The article estimates that the portion captured by the farmer in conventional agriculture is 7.8 cents, which is what I assume for the conventional farms. Direct marketing allows a small vegetable farm like Singing Frogs to capture the farmer, wholesaler, retailer, food processor portions (amounting to 45.5 cents) as well as a fraction of the food service provider portion of 36.3 cents. So for Singing Frogs Farm I estimate the portion of the consumer dollar captured at 62 cents. Dividing that by 7.8 cents for a conventional farm and rounding to the nearest integer I arrive at my estimate of the **FarmerMarketShareRatio** of 8.

Note: the Post article includes seven other much smaller portions amounting to a total of 19.3 cents which I assume to be the same for both types of farms. They won't be exactly the same but remember this is a very rough estimate.

Plugging these estimates into the above equation gives a **YieldPerAcreRatio** = 7.

My rough estimate of the yield per acre arising from Singing Frogs Farm's intensive regenerative agriculture management practices is approximately 7 times that achieved by the average California vegetable farm which is very impressive.

Further to the north in Canada's province of Quebec, Jean-Martin Fortier has achieved very similar results on an internationally recognized intensive micro-farm of 1.5 acres named Les Jardins de la Grelinette that he founded with his wife Maud-Hélène Desroches in 2004. Jean-Marin is the author of *The Market Gardener*, a very successful handbook[203] for small-scale organic farming with a theme of grow better not bigger. Their growing season is very short in Quebec but they have achieved considerable success in season extension using greenhouses following the works of Eliot Coleman and others.

Eliot Coleman's *The Winter Harvest Handbook* is a comprehensive manual on how to harvest vegetables all-year round using deep-organic techniques in even the coldest climates with unheated or minimally heated movable plastic greenhouses. For reasons that would be considered very up-to date today, much of the inspiration for Eliot's experiments in all year market gardening comes from the Parisian market gardeners of 150 years ago. They achieved all year locally grown food with excellent variety and amazing productivity, 4 to 8 harvests per year. Heat and nutrients came from composting horse manure. The average Parisian market garden at the time was between one and two acres.

Eliot Coleman and his wife and writer, Barbara Damrosch, own and operate Four Season Farm, a two acre experimental market garden on a 14 acre farm they cleared from a spruce/fir forest back in 1968. The farm is situated on a stony peninsula sticking out into Maine's Penobscot Bay in a climate with long cold winters. They have 12,000 sq. ft of greenhouse space and plan on getting 3 crops per year for each unheated sq. ft. One 3000 sq. ft. greenhouse is kept at a temperature just above freezing for the less hardy salad crops. Altogether, they grow 55 different vegetables and achieve a gross income from field crops and greenhouse crops together of over $100,000 per acre.

Here is my favorite Eliot Coleman quote, "Biotechnology companies think they have the power to remake parts of nature they don't understand. However, if they understood them, they would realize they don't need remaking. It is just our human relationship with the natural world that needs remaking."

With a perspective going back to 1965, Eliot provides valuable insights into some current organic farming practices that are shared on the farm website: https://fourseasonfarm.com.

On a still smaller scale is La Ferme du Bec Hellouin, developed by Perrine and Charles Heuvé-Gruyer in Normandy, France. *Miraculous Abundance* is the tale[204] of the couple's evolution from creating a farm to sustain their family to delving into an experiment in how to grow the most food possible, in the most ecological way possible. Their goal is to create a farm model to carry us into a post-carbon future when energy is scarcer, and localization is a must. Scientific studies[205] carried out at their farm by the Institut National de la Recherche Agronomique (INRA), AgroParisTech, the University of Gembloux (Belgium), and other organizations, validated the exceptional and sustainable productivity of the farm. In 2015, 1,000 square meters (only 0.1 hectares) of farm land at the Bec Hellouin yielded products with a market value of €55,000. That is the equivalent of €550,000 per hectare = $270,000 USD per acre. The average yearly production in France is €30,000 per hectare = $36,405 USD per hectare =$14,700 USD per acre. Their latest book *Vivre avec la terre* ('Living with the Earth'), published in May 2019, is their founding work on ecoculture. An English version is in the works.

The general theme of this chapter strongly suggests that intensively managed small farms are much more productive in yield per acre than large conventional farms. This finding is in line with the not so well known inverse relationship between farm size and productivity[206], when productivity is measured as physical yield per unit area. Results for the country of Brazil indicate the yield per acre of the smallest farms is up to 13.5 times the yield per acre of the largest farms[206]. Small farms are no doubt more labour intensive but when they achieve much higher yields without requiring expensive, environmental disturbing inputs, such as toxic chemicals and GMO seeds, then that is an added advantage.

On a per acre basis, these biointensive small farms are achieving much larger gross revenues and net revenues in the range of 40% to 85% that go directly to the farmer and farm workers. As we saw in Chapter 3, Canadian farmers practising conventional industrial agriculture net on average only 3% of the gross revenue generated by their farms prior to subsidies and carry a huge debt load. The average French farm is approximately 135 hectares and provides an income for one farmer. Based on the examples of the very small intensively managed farms described in this chapter, one hectare alone can provide an income for multiple farm workers. On this basis, as well as for environmental and social sustainability considerations, conventional industrial agriculture on large farms is possibly the worst way of attempting to feed the world.

Chapter summary:
- For environmental, social, and economic sustainability, biointensive small farms practising regenerative agriculture or agroecology offer the best hope for our future.
- On a per acre basis, they are capable of sustainably achieving much higher yields, greater employment, higher gross and net revenues for the farmer and farm workers, while simultaneously increasing biodiversity and soil fertility.

PART 4
IMPACT OF FOOD PRODUCTION ON HUMAN HEALTH

23. The hidden microbial world

Soil food web and soil microbiome

Healthy soil contains approximately 15 tonnes of biological organisms per hectare, equivalent to the weight of 20 cows[122]. This is now commonly referred to as the soil food web, a huge diversity of living things including the microbial world, creatures much too small to be seen with the naked eye. Some of the important microbes in the soil are bacteria and fungi, and a hierarchy of predators. All the genetic material these microbes contain is referred to as the soil microbiome. This is nature's impressive underground work force that we can harness if we stop focusing on killing it the way we have been doing with industrial agriculture. For too long, we have been unaware of the essential role these soil microbes play, just as we have been unaware of the importance of the vast array of microbes that live in and on every human.

Recently, a new census was carried out of the estimated 550 Billion tonnes of carbon that is distributed among all of the kingdoms of life on Earth[123]. Based on Figure 1 of reference [123], plants account for 82.5% of the total carbon based life. Remarkably, microbes account for 17.1%, and all the rest (including insects, fish, molluscs, livestock, humans, other animals and birds), account for only 0.4%.

Human microbes and the human microbiome

It turns out we have a vast array of microbes, resident in our bodies and on our skin. While bacteria are thought to be a big player, they also include fungi, archaea, viruses, and others. The biggest concentration is about 3 to 4 pounds of microbes living in our gut. We are dependent on these microbes to help digest our food, produce certain vitamins, make neurotransmitters needed by our brain, regulate our immune system, and keep us healthy by protecting us against disease-causing bacteria.

It wasn't that long ago that we thought almost all bacteria were pathogens. In reality the pathogens are kept in check by the much greater number of beneficial bacteria. Every day we are learning more about how our health depends on our microbes. It is currently the hottest topic in medicine. Our microbes even outnumber the cells in our body. Current estimates of the number of human microbes range from 1.3 to 10 times the number of human cells. You might think we are human because of our DNA but our microbial DNA has from 100 – 1000 times more genes[124]. So open your mind and behold our new idea of you.

In the next seven chapters we will explore the human health consequences of the way we grow and produce our food.

24. Effect of antibiotics on our microbiome

In 2002, French scientist Jean-Martin Bach MD DSc, drew attention to the rapid rise of many chronic diseases[125] that were occurring at the same time that modern medicine was bringing about a rapid decline of infectious diseases. Among his examples of chronic disease were asthma, Crohn's, diabetes and multiple sclerosis. Back then we didn't know about all the beneficial bacteria living in our gut. Now we know, but in the meantime it appears that antibiotics have been destroying essential human microbes[126]. Since the 1950s we have been adding antibiotics to livestock feed because it accelerates animal growth and is cheaper than other supplements[127]. When antibiotics are fed to animals they are usually referred to as antimicrobial drugs. Antibiotic resistant strains have become very common in medicine so how significant are the antibiotics fed to animals.

By 2015, the total quantity of antibiotics sold for this purpose in the U.S. reached a peak of 15.6 million kg[230]. Of that 9.7 million kg are considered "medically important" in human medical therapy. The remainder were classed as "not currently medically important." Starting with the 2016 report, that category was changed to "not medically important." For comparison the quantity of antibiotics sold for use in human medicine[231] has been relatively constant for more than a decade in the U.S. coming in at 3.4 million kg in 2017. The medically important antibiotics fed to animals declined to a minimum of 5.7 million kg in 2017 when new rules were implemented that banned the use of medically important antibiotics for growth promotion and required veterinary oversight for using antibiotics in water and feed. However in the last two years the value has been creeping upwards[232] reaching 6.2 million kg in 2019.

All this pales compared to the effect of glyphosate, the most widely used agricultural pesticide in the world. Pesticides include insecticides, herbicides, fungicides, bactericides, rodenticides, and larvicides. Glyphosate is generally classified as a herbicide. Glyphosate is the active ingredient in Roundup® and many other herbicides. It is not well known that glyphosate is also a very effective antibiotic[27][128]. When glyphosate was first introduced in the 1970s it was promoted as perfectly safe for mammals. Back then we didn't know about our gut microbiome. In 2017, the U.S. alone sprayed 127 million kg[233] of glyphosate which is a staggering 37 times the 3.4 million kg of antibiotics used in the U.S. for human medicine. We are now in the midst of a first world epidemic of chronic diseases which is predicted to bankrupt the U.S. economy within only 17 years[133][134][135][136]. The threat of antibiotic overuse goes far beyond resistant infections.

Glyphosate is much better known as a broad-spectrum herbicide that is used in more than 700 different products from agriculture and forestry to home use[129], to kill anything deemed to be a weed. Glyphosate is a potent chemical that kills almost any plant it is sprayed on that hasn't been genetically engineered[133] to be resistant. The pace of spraying greatly increased when genetically engineered glyphosate-tolerant crops were introduced in 1996. Then, instead of needing to be sprayed selectively which is labour intensive, it became possible and convenient to spray the whole field with the herbicide, including the cash crop. The amount used has been doubling every 6 years. Worldwide we are now about 2 billion kg per year. This is leading to a rise in glyphosate resistant weeds which is posing a serious problem for farmers relying on this technology[137]. Actually, herbicides do not induce resistance in weed species, rather they simply select for resistant individuals that naturally occur within the weed population. Once a resistant plant has been selected, repeated use of a herbicide over multiple generations allows resistant plants to proliferate as more susceptible plants are eliminated. Efforts to overcome the resistant plants by introducing other herbicides like dicamba have caused other serious problems for farmers[138]. On Feb. 16, 2020, a Missouri jury awarded a peach grower $265 million from Bayer and BASF in a dicamba-based weedkiller lawsuit[139]. On June 4, 2020, the U.S. Ninth Circuit Court of Appeals outlawed dicamba[140]. As a result of the Court's decision, EPA issued cancellation orders outlining limited circumstances under which existing stocks of the three affected products could be distributed and used until July 31, 2020. Then on Oct. 27, 2020, the EPA announced that they believe new analyses address the Court's concerns in regard to EPA's 2018 dicamba registrations. As a result the EPA is approving new five-year registrations for two dicamba products and extending the registration of an additional dicamba product[262].

Monsanto has patented glyphosate as an antimicrobial (antibiotic), an anti-parasitic, an anti-malarial, and as a chelator (ties up minerals so they are not available for essential aspects of the plant biology)[27][128][141][133]. It is also water soluble and has now infiltrated every sector of the water cycle. It even comes down in the rain.

Chapter summary:
- Antibiotics are destroying essential human microbes and may be responsible for the rapid rise in chronic diseases.
- In the U.S., more antibiotics are used in agriculture to fatten up livestock than all antibiotics used in human medicine.
- Glyphosate, which is the most widely used agricultural pesticide, is a powerful antimicrobial (antibiotic). In 2018, the U.S. used a staggering 37 times more glyphosate than all the antibiotics used in human medicine.
- The rapid rise of herbicide resistant weeds is posing a serious problem for farmers relying on this technology.

25. Glyphosate sprayed on crops just before harvesting

Glyphosate, the active ingredient in Roundup® and many other herbicides, is also commonly sprayed on non GMO crops just before harvesting, to kill the crop[142]. This application stresses or kills the plants, to accelerate drying and speed the ripening of the grain immediately before harvest. It allows farmers to harvest crops such as wheat as much as two weeks earlier than they normally would, reducing the risk of a poor harvest in northern, colder regions. The killing of crops with glyphosate began in Scotland in the 1980s and became common around 1992 in wheat-growing areas of North America such as the upper Midwestern U.S. and Canadian provinces such as Saskatchewan and Manitoba[215]. In addition to wheat, glyphosate is also used as a dessicant on oats, rye, lentils, peas, flax, potatoes, buckwheat, and millet. Remember, glyphosate is a very effective antibiotic, especially for our intestinal microorganisms[173], so we are getting antibiotics freshly delivered to us in our food ready to decimate our gut bacteria. In March 2015, the World Health Organization's Agency for Research on Cancer (IARC) announced that glyphosate, the active ingredient in Monsanto's herbicide Roundup®, was probably carcinogenic to humans[250][251].

The following year, the Canadian Food Inspection Agency (CFIA) published a report[239] on glyphosate contamination in food. The CFIA found glyphosate in 30 per cent of the foods tested (including grain products and infant cereals). 3.9% of grain products were found to contain levels above Health Canada's maximum residue limits (MRLs) or "safe" limits. Overall, glyphosate residues above MRLs were found in 1.3% of samples.

In 2019, Radio-Canada reported their own analysis[240] of CFIA data. They found that overall 37% of the samples contained glyphosate residues. Even in organic products, 24% of the samples contained glyphosate residues. We will return to this in Chapter 29 where we present some of the evidence in favour of eating an organic diet. For the 20 foods with the most glyphosate, 67% of the samples contained glyphosate. It turns out that these are some of the most popular foods that include wheat products like pizza for which 90% of the samples contained glyphosate. For cookies it was 81%, and for wheat bran 96%. Some other items on this list are, oatmeal, pastas, rye flour, oats, and chickpea flour. Their results are consistent with the Canadian Environmental Defence *Whats in Your Lunch?* study[241], which found that 80% of food products that they tested contained glyphosate.

Glyphosate, leaky gut, chronic disease, and a possible new medical paradigm

Celiac disease, and, more generally, gluten intolerance, is a rapidly growing epidemic worldwide, but especially in North America and Europe, where an estimated 5% of the population now suffers from it. Celiac is a serious autoimmune disease where the ingestion of gluten can lead to damage to the small intestine. In 2013 paper, Anthony Samsel and Stephanie Seneff proposed that glyphosate, the active ingredient in the herbicide,

Roundup®, is the most important causal factor in this epidemic[238]. Because of the accelerated drying and ripening of wheat by glyphosate, the wheat gluten now comes served with a portion of glyphosate.

Increased permeability of the gastrointestinal (GI) tract or 'leaky gut', is being recognized[242][243] as an early step in the development of many acute and chronic inflammatory diseases, including celiac disease and inflammatory bowel diseases like Crohn's Disease and ulcerative colitis. The GI tract takes in food, digests it to extract and absorb energy and nutrients with the help of a vast diversity of microbes, and expels the remaining waste as feces. It is lined by a single one cell thick membrane that has the area of two tennis courts. These epithelial cells of the gut membrane are joined by collections of velcro like proteins that make up tight junctions. Tight junctions hold each intestinal cell to its neighbours and regulate the passage of nutrients, macro-molecules, and pathogens across the intestinal barrier. Tight junctions also regulate what can pass across the blood-brain barrier and hold blood vessels and kidney tubules together. Tight junctions are designed to loosen temporarily, to allow only appropriate molecules to pass through and then reform a tight binding.

Zonulin is the only human protein discovered to date that is known to reversibly regulate GI and blood-brain barrier permeability by acting as a modulator of the tight junctions. Zonulin is synthesized in intestinal and liver cells and was discovered in 2000 by Alessio Fasano and his team at the University of Maryland School of Medicine. Excess Zonulin production, causing leaky gut, can be stimulated by dietary exposure to the gluten breakdown product, gliadin, as well as glyphosate. This can allow undigested food particles to enter, triggering the immune system to attack, creating food sensitivities and long term inflammation leading to a wide range of chronic diseases. These diseases are epidemic in scale, which implies environmental triggers rather than genetic or age-related causes. See more on this topic in Chapter 27.

Three medical researchers in Virginia, John Gildea, David Roberts, and Zach Bush, have been studying the effect of gliadin and glyphosate on the tight junctions in tissue cultures of rat small intestine cell lines and human colon epithelial cell lines[244][245]. They are able to image the tight junction proteins that map out the borders of the epithelial cells. They start by developing an antibody to tight junctions proteins by exposing that protein to a rabbit. The rabbit reacts and makes antibodies to the protein. They then add a glow in the dark jelly fish protein to the antibodies. When these antibodies are added to the tissue culture each antibody protein tags a single tight junction protein. When viewed through a fluorescent microscope these glow in the dark jelly fish proteins provide a map of the locations of the tight junctions. Here is what you see when these cells are tightly velcroed together. Between two cell nuclei you see what looks like a single wall where the two cell membranes come together side-by-side. When the experimenters add typical amounts of either gliadin or glyphosate the tight bindings are disrupted and the cell walls are now separate so you see two separate walls between each pair of cell nuclei. The space between the two cell walls allows for leakage through the gut lining. Their measurements indicate that the combination of gliadin and glyphosate causes a much greater disruption of the tight bindings than either one alone.

Rather remarkably, this team of researchers has discovered a liquid mineral supplement, based on a lignite extract from fossil soil, that supports tight junctions integrity. When the lignite extract is added to cell membrane tissue cultures it restores the tight junctions even after exposing them to either gliadin or glyphosate. This improvement in tight junction integrity occurred within the first thirty minutes. Their company, Biomic Sciences, has developed a product from the fossil lignite which was initially called Restore™ but has now been renamed ION Gut Health.

What is perhaps most interesting is the way their discovery came about. Zach Bush, a triple board certified U.S. doctor, shared some of the details of the discovery in a 2018 talk to the Perfect Earth Project on how to optimize your microbial health in a toxic world[133]. In 2012, Zach was surprised to find a special type of molecules appearing in a white paper on soil science. He immediately recognized its similarity to the carbon-based redox molecules he used to make for chemotherapy, earlier in his career. The redox signaling molecules he had been working with, which were made by mitochondria, killed cancer cells as well as dictating cellular repair. The essential idea seems to be that damaged cells have the ability to repair themselves or replace themselves. Often in chronic diseases and aging, the cells are not receiving clear action-eliciting signals to do so[246]. Enter redox signaling molecules.

There appears to be an exciting new medical paradigm emerging that there is no such thing as disease, only a loss of communications. In this paradigm, what is required is to restore the communications and the body will heal itself. Zach and colleagues found out that there were millions of variants of redox molecules being made in soil by bacteria and fungi. Effectively, their goal was to test out whether adding these carbon-based redox molecules from the soil microbiome would lead to healing, which they observed as an improvement in tight junction integrity in the above in vitro experiments. Since the soils of our modern agricultural system have been badly damaged they turned to fossil soils to extract the redox molecules. Healthy soil, similar to a healthy human intestinal ecosystem, contains a vast library of nutrients, minerals, amino acids, and other complex metabolites that are released through the digestive processes from a high diversity of bacteria and fungi.

Finally, in their 2017 paper[245], Gildea and colleagues give references to other studies of substances that have shown improvements to tight junction integrity. They include natural substances like quercetin, butyrate, L-glutamine, and the probiotic Bifidobacterium, as well as the synthetic pharmaceutical zonul ininhibitor, Larazotideate.

Chapter summary:
- Glyphosate is freshly delivered in our food ready to decimate our gut bacteria and damage our gut membrane.
- One promising and rapid way gut membrane integrity can be restored is with redox signaling molecules extracted from fossil soils.
- There appears to be an exciting new medical paradigm emerging that there is no such thing as disease, only a loss of communications.

26. How does glyphosate kill

Glyphosate blocks an ancient enzyme pathway, called the shikimate pathway[144][168][129]. These enzymes are responsible for making from 4 to 6 essential amino acids that produce some of the most important compounds in food and are some of the building blocks of proteins in our body. Almost all life forms on the planet possess a shikimate enzyme pathway except for mammals. Because of this, glyphosate was thought to be perfectly safe for mammals when it was first introduced in 1974. Mammals must instead obtain these essential amino acids from their diet or from gut microbes but unfortunately plants and microbes require a functioning shikimate pathway to make these essential amino acids. We now know how important these gut microbes are to help digest our food, produce certain vitamins, make neurotransmitters needed by our brain, regulate our immune system, and keep us healthy by protecting us against disease-causing bacteria.

According to Professor Don M. Huber, glyphosate gives the plant, and many of the supporting microorganisms, the equivalent of AIDS[145][168]. It essentially shuts down plant immunity. The plant then dies from disease due to its compromised immune system. Don Huber, is Professor Emeritus of plant pathology at Purdue University. For many years he has advised U.S. agencies on bio-terrorism and biological warfare. Many of his invited talks and interviews are available on YouTube.

We were always assured that glyphosate biodegraded rapidly, but this appears not to be the case. In 1996, New York's attorney general sued Monsanto over the company's use of "false and misleading advertising" about Roundup®. Monsanto had made claims that its spray-on glyphosate based herbicides, including Roundup®, were safer than table salt and "practically non-toxic" to mammals, birds, and fish, "environmentally friendly", and "biodegradable". Citing avoidance of costly litigation, Monsanto settled the case, admitting no wrongdoing, and agreeing to remove the offending advertising claims in New York State[235], and agreed to pay $50,000 toward New York's costs of pursuing the case.

In France environmental and consumer rights campaigners brought a case in 2001 accusing Monsanto of presenting Roundup® as "biodegradable" and claiming that it "left the soil clean" after use. Glyphosate, Roundup®'s main ingredient, was classed by the European Union as "toxic for aquatic organisms" and "dangerous for the environment". In January 2007, Monsanto was convicted of false advertising and fined 15,000 euros. On 15 October 2009, France's Supreme Court upheld two earlier convictions against Monsanto[236].

According to Don Huber, glyphosate is a very robust chemical[145]. When it gets into soil the majority of it binds to soil minerals and degrades only very slowly, on time scales ranging from 220 days to 6 years[164][165][166]. The remaining glyphosate, that is not bound up, is broken down quickly on a time scale of the order of 10 days[164] to a compound called AMPA[167]. This 10 day value is typical of the time scale that is widely reported for the half-life of glyphosate. Unfortunately, AMPA is just as toxic as glyphosate[144] and the carbon phosphate lyase enzyme, required to break down AMPA, is extremely rare in our soils[145]. Thus, the degradation of all toxic stages in the breakdown of glyphosate, all the way down to CO_2, phosphorus, and water, can take a very long time.

Plants are normally a source of organic compounds called alkaloids. Alkaloids have a wide range of beneficial activities for human health including anti-parasitic, anti-diabetic, anti-cancer, anti-hypertensive, anti-mood disorder, anti-depressant, anti-asthma, anti-eczema, antimalarial, and analgesic. About 20% of plant species accumulate alkaloids, which are mostly derived from amino acids like phenylalanine, tyrosine, tryptophan, and lysine. The first three of these amino acids are not produced when the shikimate pathway is knocked out by glyphosate. According to Zach Bush, if we added a chemical to our food chain that wipes out all the production of alkaloids in our food, we would have just lost the medicinal quality of our food that has existed for thousands and thousands of years[133][144][234].

Chapter summary:
- Glyphosate blocks the shikimate enzyme pathway that produces certain essential amino acids in plants and microbes. Plants and microbe transform these amino acids into many compounds beneficial to our health.
- Glyphosate and its toxic breakdown product AMPA can be very long lived in our soils.
- Glyphosate in our food chain can knock out the medicinal quality of our food.

27. The rise of chronic diseases

In Chapters 24 to 26, I outlined some of the negative biological effects of glyphosate on plants, microbes, and the human gut. It is natural to ask the question is there a correlation between the amount of glyphosate being sprayed and the incidence and prevalence of chronic diseases that exploded in the 1990s especially since the pace of spraying greatly increased when genetically engineered (GE) glyphosate tolerant crops were introduced in 1996.

In 2014, Nancy L. Swanson and three other colleagues[135] searched U.S. government databases for GE crop data, glyphosate application data and disease epidemiological data. Correlation analyses were then performed between the incidence or prevalence of 22 chronic diseases versus glyphosate application as well as disease versus the percentage of GE crops being used. Incidentally, their paper references additional negative biological effects of glyphosate that are not discussed in this book.

For each disease they computed the degree of correlation, called an R value, which is a number between 0 and ±1, where R = 0 means no correlation and R = +1 corresponds to perfect positive correlation. R = -1 corresponds to a perfect negative correlation. For example, if there is a perfect positive correlation between the amount of glyphosate sprayed and the incidence or prevalence of a chronic disease then every time the amount of glyphosate sprayed increases there will be a corresponding increase in the disease. Similarly if the amount glyphosate sprayed decreases this will result in a decrease in the disease. In other words they are moving in the same direction. In the case of a perfect negative correlation (R = -1), they are always moving in the opposite direction. A negative correlation is often referred to as an inverse correlation.

Swanson and colleagues also calculated the significance of each correlation, meaning how confident they are in rejecting the hypothesis that there is no correlation (R = 0). In statistics, the P value is the measure of significance. The smaller the P value the greater the significance of the correlation but the P value alone says nothing about the strength of the correlation which is why we need the R value as well. If the P value is less than 0.01 the correlation is described as highly significant. For all the correlations reported in the following two paragraphs the P values were less than or equal to 0.0004, so highly significant.

They found the correlation was highly significant between glyphosate applications and hypertension (R=0.923), stroke (R=0.925), diabetes prevalence (R=0.971), diabetes incidence (R=0.935), obesity (R=0.962), lipoprotein metabolism disorder (R=0.973), alzheimer's (R=0.917), senile dementia (R=0.994), parkinson's (R=0.875), multiple sclerosis (R=0.828), autism (R=0.989), inflammatory bowel disease (R=0.938), intestinal infections (R=0.974), end stage renal disease (R=0.975), acute kidney failure (R=0.978), cancers of the thyroid (R=0.988), liver (R=0.960), bladder (R=0.981), pancreas (R=0.918), kidney (R=0.973), and myeloid leukemia (R=0.878).

They also found the correlation was highly significant between the percentage of GE corn and soy planted in the U.S. and hypertension (R=0.961), stroke (R=0.983), diabetes prevalence (R=0.983), diabetes incidence (R=0.955), obesity (R=0.962), lipoprotein metabolism disorder (R=0.955), alzheimer's (R=0.937), parkinson's

(R=0.952), multiple sclerosis (R=0.876), hepatitis C (R=0.946), end stage renal disease (R=0.958), acute kidney failure (R=0.967), cancers of the thyroid (R=0.938), liver (R=0.911), bladder (R=0.945), pancreas (R=0.841), kidney (R=0.940), and myeloid leukemia (R=0.889).

In commenting on their results they point out that some of the disease plots show a significant linear rise that began prior to 1990 which they allow for. Others show smaller peaks in the 1980s, then a decline followed by the much bigger rise in the 1990s and they conclude there are multiple factors involved. Though the data for glyphosate are only available beginning in 1990, glyphosate was first introduced in the marketplace in 1974.

Altogether they have data for 22 diseases, all with a high degree of correlation and very high significance. Although correlation does not necessarily mean causation, they argue that since many of these diseases can be directly linked to glyphosate and in some cases to GE crops, via known biological effects, it would be imprudent not to consider causation as a plausible explanation. They consider it highly unlikely that all of these can be random coincidence. Ruling out coincidence, they are left with three options:

1. There is a direct cause and effect relationship
2. The relationship may be caused by a third variable
3. The relationship may be caused by complex interactions of several variables

Some of the building blocks that are being deleted in the human gut, because of the effect of glyphosate on our microbes, include phenylalanine, tyrosine, and tryptophan[144][128]. These are needed for the production of neurotransmitters like dopamine, serotonin and melatonin. We are seeing a rapid rise in all kinds of neural disorders. It turns out that our gut microbes are responsible for the production of 50% of our dopamine and 90% of the serotonin[145][146]. Deficient dopamine is associated with a decreased ability to feel pleasure and with low drive and motivation. Serotonin is responsible for our sense of happiness and well being.

According to Don Huber, there is an epidemic of 32 diseases that are related to glyphosate shutting down our access to key nutrients[145]. Sperm counts in men from America, Europe, Australia and New Zealand have dropped by more than 50 percent in less than 40 years[130][131]. By 2011, close to 50% of U.S. children had a chronic disease[132]. What does this mean for social security and pension programs that depend on a base of healthy young working people making contributions to these programs?

According to the Centers for Disease Control, in the U.S. alone, chronic diseases are the leading cause of death and disability and the leading driver of the $3.8 Trillion in annual health care costs[248], accounting for 86% of aggregate healthcare spending[261]. Chronic diseases are now the dominant factor in public health. To a good approximation we can say that as chronic diseases increase, public health decreases. Thus the explosive rise of chronic diseases corresponds to an effective collapse of public health. This systemic collapse of public health is not being adequately recognized by the public. It is essential to shift the focus from managing symptoms, which is hugely profitable for the pharmaceutical industry, to identifying systemic root causes to have any chance of regenerating human health.

It is indeed unfortunate that so many existential crises like climate change, the looming collapse of agriculture, the collapse of human health, and the collapse of biodiversity are all coming to a head. It seems clear that they are all interrelated and represent a serious deficiency in human ability to plan holistically. We excel at short term maximization of a few variables and act like we are still largely unaware of the mounting externalized costs to the environment, society, and human health. This book is my attempt to highlight the web of interconnections and likely root causes and provide a compass forward.

Chapter summary:
- We are witnessing an explosive growth of chronic diseases that are epidemic in scale, which implies environmental triggers rather than genetic or age-related causes. This translates into a systemic collapse of public health.
- The agricultural herbicide, glyphosate, and GE crops are prime suspects.
- To have any chance of regenerating human health, it is essential we shift the focus from managing symptoms to identifying systemic root causes as part of a holistic framework.

28. Processed food diseases

There are other changes that have occurred in food production that are also thought to play an important role in the growth of some chronic diseases. Virtually every processed food is now laced with added sugar on purpose because it is addictive[148][149]. The development of high fructose corn syrup (HFCS) significantly expanded this practice because HFCS is so cheap due to farm subsidies for corn. Most HFCS is composed of 42% glucose and 55% fructose and so it is quite similar to sucrose (ordinary sugar) which is 50% glucose and 50% fructose. The processed food industry got under way around 1965 but HFCS was only introduced into the American market in 1975[150]. Coincidentally, the potent herbicide, glyphosate, came to market in 1974, almost at exactly the same time.

There is an enzyme in our gut that quickly splits sugar or HFCS into their two components, glucous and fructose. Now glucose is used by all the organs in the body including the liver. Every cell in the body can use glucose and so can every other living thing because glucose is the energy of life. About 20% of the glucose reaches our liver where some of it is converted to glycogen which is a non-toxic storage form of glucose. Prof. Robert Lustig explains that the situation is very different for fructose because only the liver can metabolize it, and that fructose results in many negative impacts for human health[148]. Robert Lustig, MD, is Professor Emeritus of Pediatrics, Division of Endocrinology and Institute for Health Policy Studies, University of California at San Francisco.

Among other things fructose leads to insulin resistance so cells in your muscles, body fat, and liver, start ignoring the signal that the hormone insulin is trying to send out - which is to grab glucose out of the bloodstream and put it into our cells. Over time, this sends up your blood sugar levels. That can set you up for type 2 diabetes, as well as heart disease. Insulin resistance causes your pancreas to make more insulin in order to help the liver do its job. The high insulin goes to the brain where it blocks another hormone called leptin which is vital in the regulation of appetite. So what does that do? It makes you think you are starving, leading to the consumption of more fructose. So now you've got a positive feedback effect causing damage to the liver, damage to the pancreas, and eventually damage to the brain. Robert Lustig points out that a calorie of fructose does not behave like a calorie of glucose in the human body. This is very different from the claim of food industry adds, that all calories are equal.

Wait a minute, don't fruit and some root vegetables contain fructose as well? Fruit contains both fructose and glucose. The proportions of each vary, but most fruits are about half glucose and half fructose. In fruit and root vegetables, the sugar is packaged with fibre. According to Robert Lustig the antidote for fructose is fibre. Therefore, it is wise to avoid those foods that contain ingredients that have had the fibre removed. This would include processed foods, sugary foods, and even fruit juices. Choose to eat whole foods instead. Up to 31% of the total fibre in a vegetable can be found in its skin along with a lot of the vitamins and antioxidants[151].

An historical analysis of sugar industry documents, published in 2016, found that the sugar industry sponsored a research program in the 1960s and 1970s that successfully cast doubts about the hazards of sucrose while promoting fat as the dietary culprit in coronary heart disease[152]. Recent analysis shows just the opposite to be true[153]. High carbohydrate intake are associated with higher risk of total mortality, whereas total fat and individual types of fat are related to lower total mortality. Clearly, policy making committees should give less weight to food industry funded studies.

Processed fructose, mostly in the form of corn syrup, has become a major contributor to the $3.8 trillion health care budget[261] in the United States. According to Robert Lustig, there are clear data linking sugar consumption to de novo lipogenesis - a disease process associated with fat accumulation in the liver that causes insulin resistance and leads to something called metabolic syndrome and associated diseases[148]. They include Type 2 diabetes, hypertension, lipid problems, cardiovascular disease, cancer and dementia. Even though fructose damages the liver the FDA can't and won't regulate it. The FDA only regulates acute toxins, not a chronic toxin like fructose. The liver doesn't get sick after one fructose meal but more like a thousand fructose meals[150].

According to Lustig[154], "We now have the smoking gun. We have the mechanism by which this occurs. In fact, our paper just appeared in *Gastroenterology*[156], which demonstrates that if you take sugar out of the diet of children with metabolic syndrome and substitute starch – calorie for calorie exchange, glucose for fructose exchange, no change in calories and no change in weight – in 10 days, you can reverse metabolic syndrome. You can reverse the insulin resistance. You can reverse the liver fat. You can reverse the burden on the pancreas. Basically, all of the metabolic perturbations go away. This is the smoking gun." On this basis fructose ingestion interferes with obesity intervention.

An article in European Scientist[156] entitled "Covid 19 and the elephant in the room" by Dr. Aseem Malhotra, points out that obesity, type 2 diabetes and a cluster of risk factors all linked to poor diet is the root cause behind increased death rates from COVID-19. Dr. Malhotra is an NHS Consultant Cardiologist and Professor of Evidence Based Medicine. Drawing on UK data, Dr. Malhotra notes that 72.7% of patients admitted to ICU are overweight or obese. A recent commentary in the journal Nature[157] indicates that patients with type 2 diabetes and metabolic syndrome might have up to a ten-times greater risk of death when they contract COVID-19. Referring to U.S. data where obesity levels are similar to the UK, Dr. Malhotra notes that only 12.2 % of American adults are considered metabolically healthy, with less than a third of normal weight people also in this category. The elephant in the room is that the baseline general health in many western populations is already in a horrendous state to begin with. COVID-19 is literally unmasking the poor underlying health of the population. International experts call for immediate update on public health messaging to eat whole nutritious food to rapidly reduce risk of COVID-19 complications and potentially save hundreds and thousands of lives.

Professor Robert Lustig is author of New York Times best selling book *Fat Chance: The Bitter Truth About Sugar*, as well as *The Hacking of The American Mind: The Science Behind the Corporate Takeover of Our Bodies and Brains*.

According to Lustig, the dopamine pathway in the brain, is the reward pathway. It is the same no matter what your source of pleasure is. It can be a substance, such as nicotine, alcohol, heroin, cocaine, sugar-containing food, or it can be behavior, such as internet, shopping or porn. The bottom line is that the dopamine pathway is exactly the same. Sugar is considered the cheapest gateway drug to other addictions.

Because dopamine is an excitatory neurotransmitter, when it is released and the neuron on the other side accepts that signal, it can cause damage. Over time, excitatory neurotransmitters can cause cell death, so the neuron has a method for dealing with it. It downregulates the receptors, so there are fewer of them resulting in less damage. When you get a hit, you get a rush. The receptors go down. The next time, you need a bigger hit to get the same rush. The receptors go down and down, until finally, with a huge hit you get nothing.

One problem is that as dopamine goes up, serotonin goes down (dopamine downregulates serotonin) and the less happy you feel. Also, a lack of sleep or stress causes cortisol to increase which also downregulates serotonin. So excess dopamine and cortisol are a recipe for both addiction and depression. Serotonin itself is not an excitatory neurotransmitter so there's no such thing as overdosing on too much happiness. Robert Lustig goes through the fascinating science of these neurotransmitters in his book, *The Hacking of the American Mind*, and in many of his presentations. He offers the four C's to up your serotonin, tamp down your dopamine, and lower your cortisol. The four C's are Connect (face-to-face), Contribute, Cope, and Cook[158].

> **Chapter summary:**
> - Foods high in fructose and low in fibre damage the liver, pancreas, and brain; a big factor in chronic diseases.
> - Corporations engineer our cravings to takeover our bodies and brains causing real happiness to elude many of us.

29. Glyphosate in breast milk and pesticide safety issue

In 2014, a non profit consumer education group, "Moms Across NA," raised money to test for glyphosate in breast milk of ten of their members who were trying to stay away from Roundup® and glyphosate[159]. Five had detectable levels and sadly three had levels that were 760 to 1600 times levels allowed in EU drinking water. The study also found that urine from American mothers contained levels of glyphosate ten times higher than the urine from European women. Note, the maximum allowable level of glyphosate in Canadian drinking water[160] is 2800 times higher than EU levels[162]. The U.S. Environmental Protection Agency (EPA) maximum contaminant level for glyphosate in drinking water[161] is 7000 times the EU level.

The following year in March 2015, the World Health Organization's Agency for Research on Cancer (IARC) announced that glyphosate, the active ingredient in Monsanto's herbicide Roundup®, was probably carcinogenic to humans[250][251]. Two other commonly used pesticides, malathion and diazinon were also declared to be probably carcinogenic to humans by IARC in 2015. Recently, two peer-reviewed-studies[174][175] showed that an organic diet intervention significantly reduced urinary pesticide levels in U.S. children and adults for 14 pesticide metabolites and parent compounds including glyphosate. They studied pesticide levels in four American families for six days on a non-organic diet and six days on a completely organic diet. Switching to an organic diet decreased levels of glyphosate by 70% in just six days. The biggest reductions observed were 95% for a metabolite of malathion, 87% for clothianidin, and 60.7% for a metabolite of chlorpyrifos. **When you choose organically-grown products, you're guaranteed they were not grown with the roughly 900 synthetic pesticides allowed in non-organic agriculture**[176].

> **What are pesticide metabolites?** In some cases, alterations to the active ingredient of a pesticide are required to confer toxic effects. For instance, most systemic insecticides are converted into metabolites that are more water soluble and are more active against insect pests[132b]. A systemic pesticide is any pesticide that is absorbed into a plant and distributed throughout its tissues, reaching the plant's stem, leaves, roots, and any fruits or flowers.

Chlorpyrifos is an organophosphate insecticide used on a large number of crops worldwide. It belongs to a class of chemicals developed as a nerve gas and is now found in food, air and drinking water. Human and animal studies show that it damages the brain and reduces I.Q.s while causing tremors among children[178]. The U.S. EPA was preparing to ban it for agricultural and outdoor use in 2017, but then the Trump administration rejected the ban[258]. According to the European Food Safety Authority there is no safe exposure to chlorpyrifos[179]. At the time of writing, all four of the pesticides mentioned above are registered for use in Canada by the Health Canada Pest Management Regulatory Agency[180].

In 2020, I interviewed Simon Fraser University Health Science Professor Bruce Lanphear MD, about his perspective on agricultural pesticides. His research helps to quantify and prevent disease arising from toxic chemicals and pollution. Here are several quotes from Bruce Lanphear.

"We've made so many mistakes over the past century.
- First, we liberally sprayed our orchards and farms with lead arsenate until it was found to be toxic.
- Next, we doused them with DDT until it was found to be dangerous.
- Then we sprayed our farms and gardens with organo-phosphate pesticides. Oops, they are poisonous!
- Today, widespread applications of pyrethroids, neonicotinoids, and glyphosate are growing even as new evidence shows they are toxic too."

"In the meantime, we've learned that we don't need pesticides to feed the world or to make our gardens pretty." "Why do we keep using toxic chemicals on our farms and in our gardens? It isn't enough for scientists to speak out, we need the public to stand up and demand change. Our leaders need our help to protect us and our children from toxic chemicals."

Here is another quote from Professor Don Huber pertaining to glyphosate[168]. "The U.S. EPA is repeatedly approached by the companies that say that we have to increase the amount of glyphosate that we can have in our food, so we can have a safe product – not based on science but based on how much chemical is actually in our food. May 1, they just doubled the amount of glyphosate that can be in our food. In soybean oil, you can have 40 parts per million. Dr. Monika Krüger's research at the Leipzig University shows that a tenth of a part per million is all that it takes to kill your lactobacillus, bifidobacterium, and enterococcus faecalis. Soybean oil is now allowed to contain a whopping 400 times the limit at which it can impact your health."

According to Don Huber unless we stop using glyphosate right away we are doomed! Recently, Zach Bush, a triple board certified U.S. doctor, weighed in on the topic. He spent 20 years working with cancer patients using radiation, chemotherapy and surgery until he realized that cancer had gone from a genetic disease (according to the prevailing view of the medical profession) to an epidemic along with about 30 other chronic diseases. According to Zach Bush, we are dealing with one of the most extraordinary explosions of disease this human planet has ever seen[133][134]. We are killing our microbiome especially with our massive use of agricultural antibiotics.

Chapter summary:

- In 2014, glyphosate was detected in the breast milk of five out of a group of ten North American mothers who were trying to stay away from Roundup® and glyphosate. In two cases the level greatly exceeded the allowed levels in EU drinking water by as much as 1600 times.
- The U.S. EPA maximum contaminant level for glyphosate in drinking water is 7000 times the EU level.
- The Canada Government maximum allowable level for glyphosate in drinking water is 2800 times the EU level.
- An organic diet can significantly reduce urinary pesticide levels.
- When you choose organically-grown products, you're guaranteed they were not grown with the roughly 900 synthetic pesticides allowed in non-organic agriculture.
- History shows we are not adequately testing agricultural chemicals before inflicting them on our children and the environment.

30. Fake science on trial in the courts

On Aug. 10, 2018, a San Francisco jury found in favour of a school groundskeeper dying of cancer, whose lawyers argued that a weed killer made by the agribusiness giant Monsanto likely caused his disease. The jury awarded the plaintiff, Dewayne Johnson, $289 million in damages[169][170]. According to the Associated Press, Robert Kennedy Jr., a member of Johnson's legal team, said "The jury found Monsanto acted with malice and oppression because they knew what they were doing was wrong and doing it with reckless disregard for human life." Johnson's case was particularly significant because a judge allowed his team to present scientific arguments.

Later on Oct. 22, 2018, San Francisco Superior Court Judge Suzanne Bolanos upheld the jury's verdict that found that Monsanto's weed killer caused the groundskeepers cancer, but she slashed the amount of money to be paid from $289 million to $78 million.

On Mar. 27, 2019, a federal jury awarded $80 million to another plaintiff, Edwin Hardeman, after determining that Monsanto's popular weedkiller, Roundup®, was a substantial factor in causing his cancer, and that the corporation is liable[171]. The jury in San Francisco awarded compensatory damages at $5.27 million and punitive damages of $75 million.

In May 2019, Monsanto was ordered to pay a couple $2 billion in the largest verdict yet over a cancer claim[172].

The use of glyphosate greatly increased when genetically engineered glyphosate-tolerant crops were introduced in 1996. Then, instead of needing to be sprayed selectively which is labour intensive, it became possible and convenient to spray the whole field with the herbicide, including the cash crop. Clearly it allows the chemical companies to market a great deal more product.

In April, 2017, Professor Don Huber gave a talk[27] entitled 'Disrupting the Integrity of Nature: Pesticides and Genetic Engineering', at the 35th National Pesticide Forum at the University of Minnesota. In his closing remarks he put up a slide that read, "Future historians may well look back and write about our time, not about how many pounds of pesticides we did or did not apply; but about how willing we are to sacrifice our children and jeopardize future generations with this massive experiment we call genetic engineering that is based on false promises and flawed science, just to benefit the bottom line of a commercial enterprise."

It now seems abundantly clear that genetic engineering in agriculture[173] is really about maximizing the profits of giant chemical companies and seizing control of the world's seed stocks. There is considerable momentum in the direction of genetically engineering, patenting, and controlling a much greater variety of foods not to improve human health but for increased productivity and profit. In Part 3 of this book, I discussed a very different approach to feeding the world using biointensive regenerative agricultural to work with nature and go beyond sustainability, allowing us to address a variety of existential threats facing humanity.

Chapter summary:
- May 2019, a California jury ordered Monsanto to pay a couple $2 billion in the largest verdict over a cancer claim. This is just the latest of three cases where courts have ordered the company to pay huge punitive damages for their behaviour.

31. Whose interests do health regulators serve?

All this begs the question: are government regulators working in the interest of public health or the interests of large multinational chemical companies and the processed food industry? Consider that one pesticide, glyphosate, could collapse the North America's economy in only 17 years[133]. Even Europe extended its authorization for glyphosate for an abbreviated period of five years[181]. The influence of large corporations on government policy is staggering to behold. That was the subject of the 2015 book by Steven M. Druker, *Altered Genes and Twisted Truth*. Druker is a public interest attorney who initiated a lawsuit that forced the U.S. Food and Drug Administration (FDA) to divulge its files on genetically engineered foods. This exposed how the agency had covered up the warnings of its own scientist about the risks, lied about the facts, and then ushered these foods onto the market in violation of federal law.

As recently as Jan. 11, 2019, Health Canada reaffirmed its 2017 approval of glyphosate use in Canada which it is required to do every 15 years[182]. They concluded glyphosate products pose no risk to people or the environment as long as they are properly used and labelled. Clearly, this is much easier to justify when you set the maximum allowable level for glyphosate in Canadian drinking water[160] at 2800 times higher than European level[162].

We need to dig a little deeper into the way regulatory agencies function. One opportunity to do this comes from a 2001 report issued by the Royal Society of Canada (RSC) at the request of Environment Canada, Health Canada and the Canadian Food Inspection Agency. The Panel was asked to evaluate the Canadian regulatory system and the scientific capacity needed to cope with products in the future. We are living in an age where life itself is being manipulated, picked apart, reassembled and then patented. The practitioners would like you to believe this is being carried out with great precision based on a sound understanding of the code of life. Druker's book documents in great detail how this is far from the truth. It is better described as biohacking not engineering. At a very instinctive level, there's a sense that a very robust precautionary principle needs to be in place to guide all forms of genetic engineering.

The Royal Society report[183] raised serious concerns about the undermining of the scientific basis for risk regulation in Canada due to the following factors:

- the conflict of interest created by giving to regulatory agencies the mandates both to promote the development of agricultural biotechnologies and to regulate it.
- the barriers of confidentiality that compromise the transparency and openness to scientific peer review of the science upon which regulatory decisions are based.
  (A Toronto Star article[249] of Feb. 5, 2001, *GM Food Report: Ottawa Rapped, Expert Study Considered Major Setback for Biotech Industry*, reported that Federal regulators barred even the Royal Society panel from seeing evidence that safety tests had actually been done on genetically modified foods.)

- the extensive and growing conflicts of interest within the scientific community due to entrepreneurial interests in resulting technologies and the increasing domination of the research agenda by private corporate interest.

The RSC Panel made 58 recommendations for changes to the regulatory system, many of which would have profound implications. The Panel called for a precautionary approach to the regulation of genetically modified organisms (GMOs), and made it clear that this approach should not be compromised by the commercial interests of corporations wanting to get new products to market quickly. **The RSC Panel found that, in 2001, a truly precautionary approach was not in place for GMO regulation in Canada.**

In 2004, the Canadian Biotechnology Action Network carried out a detailed analysis of all the recommendations of the Royal Society of Canada, comparing them to the Federal Government's so-called "Action Plan" developed in response[184]. Environmental non-governmental organizations and other civil society groups in Canada collaborated with independent university researchers to produce this report in order to return attention to the recommendations of the RSC Panel.

"It finds that while some progress has been made, there is still a great deal that needs to be done before Canadians have a precautionary regulatory system to protect their families and the environment from the risks of GMOs. Because of the limited progress, this report concludes, based on the rationale presented by the RSC itself in 2001 (p.225), that it is time for the Government to finally legislate mandatory labelling for all GM foods."

In a 2018 Dalhousie University study, approximately 90% of survey respondents agreed that GMO food and ingredients should be labelled on all packages[259]. BC was the highest at 91% and the prairies the lowest at 85%.

Gene-editing techniques are widely considered to offer substantial improvements, in terms of precision, over older genetic engineering techniques. Gene editing technologies, like Crispr-Cas9, alter the genome of a living species by slicing genome strands in a bid to remove undesirable traits, without necessarily inserting foreign DNA. However, a 2018 study[185] published in the scientific journal Nature found that the gene-editing technology can cause significantly greater genetic distortions than expected, with potential "pathogenic consequences". In 2016, FDA research showed that foreign DNA can become inadvertently incorporated, in this case unbeknownst to the developer[209][186]. These findings are a significant blow to the argument that gene-editing should not be subject to regulation and represent a vindication of the EU approach, which is to regulate gene-edited organisms as GMOs.

Clearly the quest for short-term profit and growth will never guarantee human survival. Many of our leaders are intelligent people but they appear locked into institutional beliefs that are leading us in the wrong direction. They are unable to remove the blinders that are preventing them from seeing the bigger picture. From a 30,000 ft view they all seem quite content reorganizing the deck chairs on the Titanic.

Chapter summary:
- All is not well in the regulation of agricultural chemicals and genetic engineering.
- A Royal Society of Canada Panel found that a truly precautionary approach was not in place for GMO regulation in Canada.
- In Canada the same agency is mandated to promote and to regulate agricultural biotechnologies, a clear conflict of interest.

PART 5
SUMMARY AND CONCLUSIONS

32. Summary

We have covered a lot of ground and perhaps it is time to summarize. I have chosen to do this in two ways. First as a one line highlight of each chapter. Then I illustrate the interconnections between the topics in two diagrams which I call alternate futures. What follows next is a table with an abbreviated chapter headings and the main point.

1. Starting point, personal health: - **my wife and I learn that our main health issues are food related**
2. Looming collapse of agriculture: - **7 tons of soil lost for every ton of food produced; largely caused by agricultural practices**
3. Green revolution: - **1950s to 1990s, plant breeding and a high input of agrochemicals yield a big increase in cereal crops**
4. We have been converting living soil to dirt: - **plowing destroys soil structure built by billions of microbes in every teaspoonful**
5. Nature's barter system: - **plants feed soil microbes carbon exudates in return for soil minerals mined & recycled by microbes**
6. More carbon in the atmosphere or in the soil: - **plowing releases stored soil carbon as CO_2, a greenhouse gas**
7. Soil carbon sponge: - **biology can transform a pile of flour into bread & a pile of dirt into soil, a water holding carbon sponge**
8. Saskatchewan no-till farmers: - **achieved a cash crop every year instead of every two years and sequestered soil carbon**
9. What are plants made of: - **plants mainly made from air and water through photosynthesis, soil mineral content < 1%**
10. Exporting soil nutrients: - **regenerative agriculture saves money and demonstrates that chemical fertilizers are not required**
11. Nutritional declines in foods: - **mounting evidence for nutrient declines of 5% to 35% and rarely 75% in the case of copper**
12. Browning of green revolution: - **N fertilizer depletes soil organic matter, stored nitrogen - leading to crop yield stagnation**
13. Nature's complexity amazing: - **herbivores, managed to mimic nature to achieve plant recovery time, reverse desertification**
14. The new scoop on methane: - **new science shows that methane from herbivores is not as big a problem we thought it was**
15. Human planning & nature's complexity: - **reductionist management is devastating natural resources that life depends upon**
16. Regenerative agriculture: - **reverses soil degradation with biology, improves soil fertility and enhances the soil carbon sponge**
17. Six principles of regenerative agriculture: - **a summary of the key principles that work in any garden or farm**
18. The global thermostat: - **why rebuilding the soil carbon sponge is essential to quickly cool the planet for our survival**
19. The ABCD of agriculture: - **the benefits of regenerative agriculture compared to conventional agriculture in one diagram**
20. One man's agriculture journey: - **from industrial agriculture to a deeply sustainable ecological approach of regeneration**
21. True cost of food: - **hidden environmental & social cost of industrial food production are at least as large as the retail cost**
22. Bountiful small regenerative farms: - **biointensive, regenerative farms provide hope for our future**
23. Hidden microbial world: - **currently the hottest topic in medicine and key to the future of regenerative sustainable agriculture**
24. Effect of antibiotics on our microbiome: - **glyphosate, the most widely used agricultural pesticide, is a powerful antimicrobial**
25. Glyphosate sprayed just before harvesting: - **glyphosate freshly delivered in our food ready to decimate our gut bacteria**
26. How does glyphosate kill: - **it blocks the shikimate enzyme pathway that makes essential amino acids in plants & microbes**
27. Explosion of chronic diseases: - **coincides with the rapid increase in glyphosate use and GE crops**
28. Processed food diseases: - **foods high in fibre-free fructose damage liver, pancreas & brain; big factor in chronic diseases**
29. Glyphosate in breast milk & pesticide safety issues: - **an organic diet can significantly reduce human urine pesticide levels**
30. Fake science on trial: - **May 2019, Monsanto ordered to pay a couple $2 billion in largest verdict over a cancer claim**
31. Public health regulators: - **the same agency promotes and regulates agricultural biotechnologies, a clear conflict of interest**

The following two diagrams (Figures 11 and 12) illustrate the links between the various topics I have discussed.

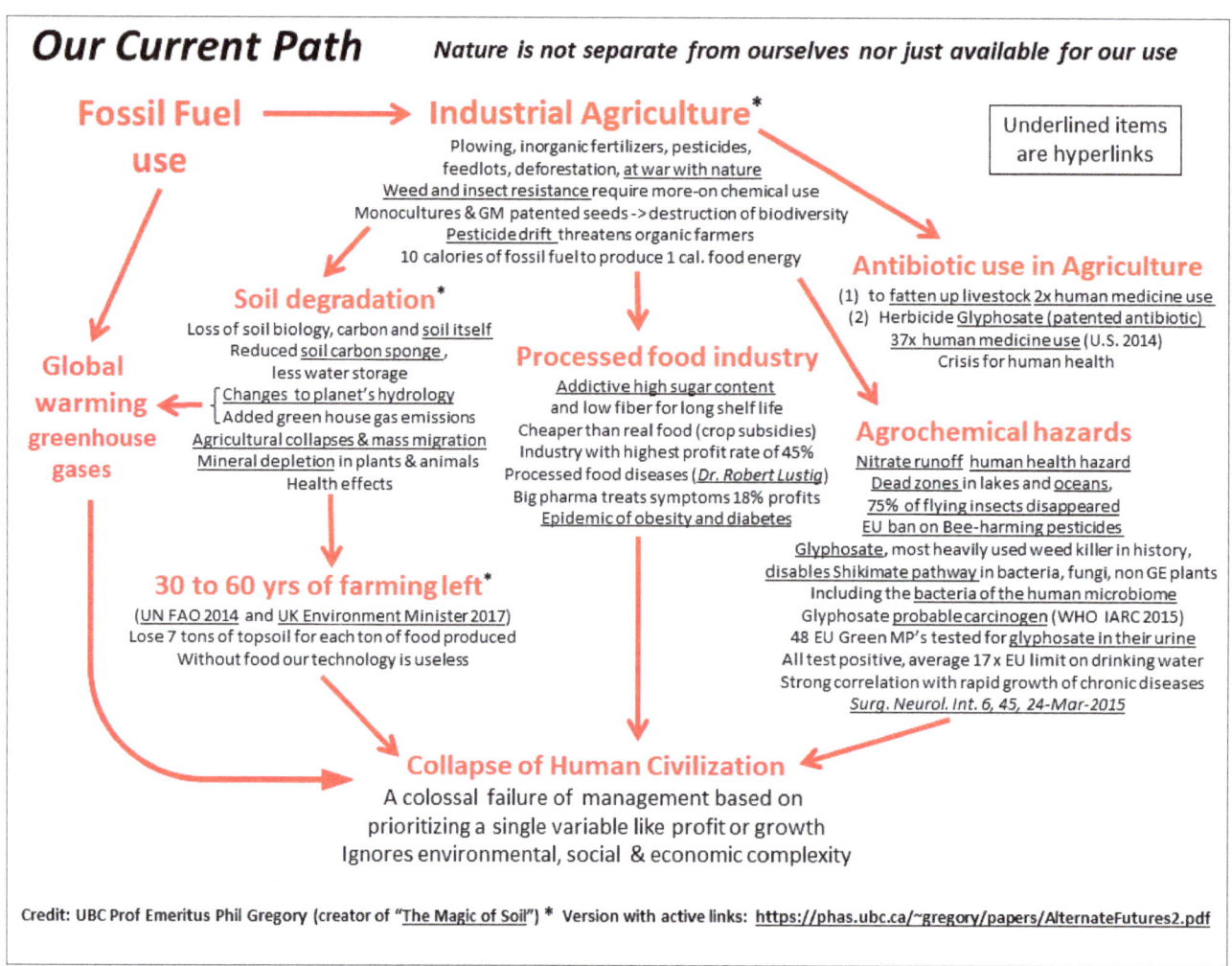

Figure 11. Some of the many interconnections between current food production, health, climate change and complexity.

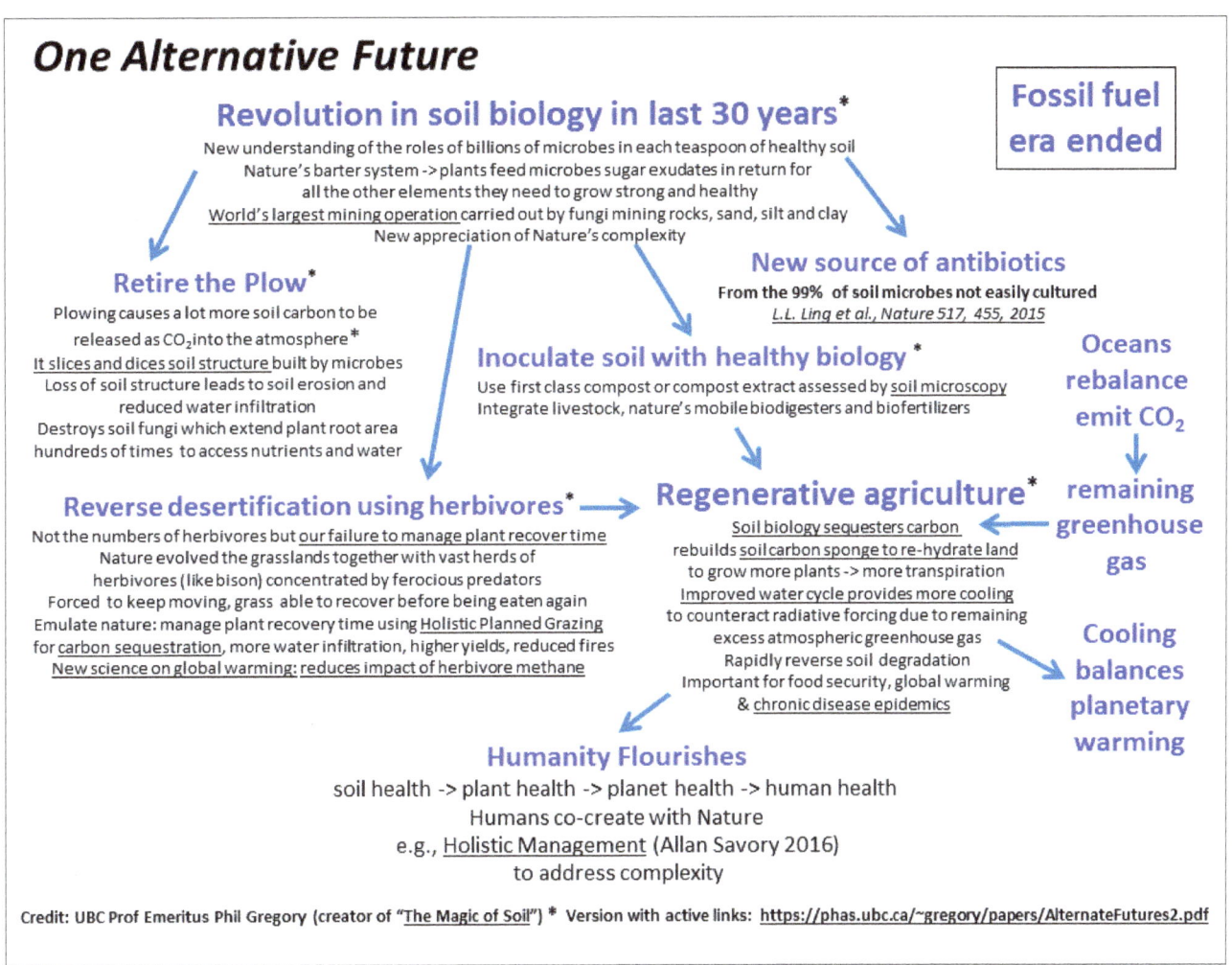

Figure 12. An alternate realignment of interconnections between food production, health, climate, and complexity based on a scientific revolution in our understanding of soil biology and nature's complexity of the last 30 years. This offers tremendous potential to transform human activity from an earthly cancer to a cooperating 'Team Nature' player. Imagine a win – win situation where working with nature actually makes our life easier and healthier.

33. So what can we do?

In my six year investigation to find out if anything could be done to tackle the looming crises in food security and global warming, I came across the epidemic of chronic diseases and their strong connection with current agricultural practices and the way food is processed. My primary focus has been on our new understanding of soil biology and nature's complexity that I now see offers tremendous potential to help tackle all three of these existential threats, if we move quickly to regenerative agriculture. **This is amazing!** This would require us to change the way we manage herbivores, stop plowing and greatly reduce many agricultural chemicals and completely eliminate others which are devastating to human health and soil health. We would be collaborating with nature, not killing it.

Of course, it is also imperative that we rapidly reduce our use of fossil fuels to prevent further warming. Fortunately, our new understanding of how to sustainably grow food is congruent with the need to reduce fossil fuels.

What was just as surprising to me is that this new knowledge is not widely appreciated. As an outsider I have been able to span many different silos of knowledge not hampered by the blinders of specialization. As an astrophysicist, it is almost as if I have come exploring from another planet. Pockets of this new understanding exist all around the world[207] and it has been happening on some pioneering farms for many decades[34]. However, it is sufficiently new for our institutions, that it is still not taught in most agricultural schools. From my observations, many university academics are still researching how to fine-tune conventional chemical-intensive agriculture which is not surprising as a lot of agricultural research is funded by chemical companies. They appear to be unaware of the revolution in soil biology as their papers don't even mention soil biology.

A good example of this thinking is reflected in the 2020 report on 'The Future of B.C.'s Food System' produced by the B.C. Food Security Task Force[208]. The focus is on the next 'Green Revolution', to supercharge the existing technology options available to farmers. Agritech is the buzzword of the report which is heavy on robotics, drones, and new ways of growing food like cellular agriculture which involves multiplying animals cells in tissue culture. The report calls for a streamlined regulatory framework to rapidly launch large scale agritech demonstration projects. The foundational belief that emerges from reading the B.C. Task Force report is that human technological silver bullet innovations are superior to anything nature has to offer. Hubris, combined with ever more powerful technology and the abandonment of the precautionary principle, is hurtling humanity to the brink of collapse.

Industrial food systems are being reinvented as the solution, despite their role in driving ecosystem disruption, climate change, and making food systems critically vulnerable to pandemics and other shocks[210]. According to Vanadana Shiva, this is fossilized thinking. Instead, working with nature's complex and highly evolved systems, including the soil microbiome, should be the hottest topic in agriculture as the human microbiome is in medicine.

- So what is to be done? At this point we need to take a closer look at how our institutions work. Humans have created institutions as an efficient way to provide public services like government, health and safety, military, legal, and education, based on the operating paradigm of the day. Unfortunately, when the paradigm changes, the reward system and dynamics of existing institutions result in the same institutions resisting any shift to a new paradigm. For example, farmers run into difficulty with crop insurance if they stop plowing or use cover crops and will often not receive a bank loan to buy seeds without using chemicals. Institutions only change when the belief system of the society has moved beyond the old paradigm and institutions realize they are out of step. In addition, institutions do not generally think holistically but are focused on protecting the funding for their own silo, profession or domain of influence. Allan Savory has given a lot of thought to this topic and his recent insights (Sept. 2018) are available on YouTube[211].
- What are some of the obstacles we face? Conventional agriculture is all about fighting nature supported by a belief system that only technology will give us the upper hand in that fight. There is little recognition that we are losing that battle and dooming humanity in the process. Here are some of the things we need to do to turn this around.
- It is imperative that we rapidly reduce our use of fossil fuels to prevent further warming.

We also need to foster:
- awareness of the ecological footprint of how our food is currently being grown[212] including the loss of topsoil, forests, soil carbon, soil microbes, soil moisture, and biodiversity.
- awareness of the mounting evidence for nutrient declines in our basic food stocks (see Chapter 11).
- awareness of the many chronic disease epidemics and their relationship to the food we are eating, how it is grown and processed. The results from one recent COVID-19 study found that of 94% of the deaths were for patients with at least one underlying health condition[213]. Nowadays the majority of health conditions are classified as chronic diseases. COVID-19 is unmasking the dire health issues in the population linked to a poor diet. The systemic weaknesses exposed by the virus will be compounded by climate change in the years to come. At the same time COVID-19 is providing a time-out to consider where life as normal is leading us.
- awareness that changes in consumer spending are happening as people decide to eat more organic real food to avoid a life of chronic disease for themselves and their children. Food distributors have already had to make the change for European markets that are demanding GMO free crops. Jeffrey Smith[214], a leading American consumer activist and educator on the dangers of glyphosate and GMOs, estimates that it will only require 5% of the population to buy organic for a tipping point to occur away from GMO/chemical producers.
- awareness that regenerative agriculture, including regenerative organic agriculture, is a viable alternative where we mimic nature and work with her incredible underground work force. This has the potential to produce more nutritious food, sequester carbon, and increase farmers' profits. Worldwide there are now many good successful examples[34][35].
- awareness that much greater yields per acre are achievable on very small biointensive regenerative farms that can support many more farmers and farm workers producing local foods in ways that are congruent with our need to get off of fossil fuels (see Chapter 22).

- awareness of the important role for herbivores, grazed holistically, to sequester more GHG than they produce and to reverse desertification, all the while serving as nature's mobile biodigesters and biofertilizers providing and distributing soil building compost (see Chapter 13).
- awareness of the many ways to rapidly cool the planet by rebuilding the soil carbon sponge with regenerative agriculture (see Chapter 18).
- awareness of the toxic effects of agricultural chemicals on human health and particularly with regard to their effect on the soil and the human microbiomes. History shows we are not adequately testing agricultural chemicals before inflicting them on our children and the environment. Many provinces, cities and communities are banning[215][216][217][218][219] the widespread use of pesticides in response to the growing epidemic of chronic diseases. These restrictions are generally limited to residential areas while farms are exempted. Some of these chemicals do not decompose in composting operations and there is an urgent need to recycle organic waste, including human waste.
- awareness of holistic management, a way to deal with the inherent complexity of the ecological systems of which we are a part. Nature needs a seat at the table. Our reductionist management is razor focused on very few variables and seems only capable of short-term benefits while building up mountains of externalized damage to the biosphere.

Other important steps:
- demand government food and health regulatory agencies be divorced from efforts to promote biotechnology and corporate interests. Many are currently faced with an extreme conflict of interest and need to focus on protecting the health of the public and the environment using a truly precautionary and transparent regulation system as recommended by Canada's Royal Society[183].
- develop accessible education material on a step-by-step approach for achieving regenerative agriculture from the garden scale to the farm scale.
- we need to develop a new economic system that prioritizes the growing of highly nutritious food as basic preventive health care and natural resource care with opportunities for young people to earn decent incomes in regenerative agriculture. According to Professor Robert Lustig, the food industry gross profit in the U.S. is dwarfed by the health care costs for 75% of chronic metabolic diseases that are preventable by an improved diet[220]. We need to take a close look at the societal math of processed food and food grown with toxic chemicals.

> "People are fed by the food industry, which pays no attention to health,
> and are treated by the health industry, which pays no attention to food."
> Wendell Berry

- One encouraging trend is the formation of alliances between farmers and doctors to develop root-cause solutions for human health and ecological health through regenerative agriculture. Two recent examples are:
    - the Farmer's Footprint, a non-profit founded by Zach Bush, MD
    - an alliance between the Rodale Institute and the Plantrician Project[96].

Based on current global trends, 6 of the top 7 causes of death in 2040 (including heart disease, stroke, Alzheimer's and diabetes) will be directly related to our lifestyle choices and diet[221].

The focus of this book has been to connect the dots between a wide range of human activities that pose an imminent existential threat, with the aim of finding a path forward for humanity. The good news is that such a path does exist which involves working closely with nature instead of our current war on nature. This requires a complete rethinking of our place in the Earth's biosphere and the need for a holistic approach to managing our affairs in recognition of the biosphere's intrinsic complexity.

The window for making the changes required is shrinking rapidly. According to a 2018 IPCC report[222], global net human-caused emissions of CO_2 would need to fall by about 45% from 2010 levels by 2030, reaching 'net zero' around 2050. Yet globally, GHG emissions are still rapidly rising, with increasingly damaging effects on the Earth's climate[223]. What is required is a very tall order but scientists are crystal clear[222] about what it will take: the next decade must see "**rapid, far-reaching and unprecedented changes in all aspects of society.**"

Consumption of solar and wind energy has increased 373% per decade, but in 2018, it was still 28 times smaller than fossil fuel consumption[223]. The International Monetary Fund makes regular estimates of worldwide subsidies to fossil fuels and in 2017 it put these subsidies at $5.2 trillion USD[224]. They partly take the form of tax breaks and outright cash grants amounting to approximately $500 billion per year. The vast majority of the IMF's $5.2 trillion subsidy tally comes from failing to price greenhouse gas emissions. In essence, the world's carbon polluters are dumping their waste into the atmosphere for free. About 87% of greenhouse gas emissions don't face any kind of carbon price at all. According to the IMF, efficient fossil fuel pricing in 2015 would have lowered global carbon emissions by 28% and fossil fuel air pollution deaths by 46%, and increased government revenue by 3.8% of GDP[224].

Even if we stopped all human emissions today the climate would continue to warm. We actually have to remove some of the excess atmospheric carbon and, as previously mentioned, regenerative agriculture has an important role to play.

It is hugely important to continue raising the level of awareness of the paradigm shift that has to occur and the time frame. In parallel we need to define and implement policies that move us in the right direction using a holistic management approach. We will not get there with existing reductionist management tools that ignore the interconnectedness of the issues.

34. How do I feel, knowing what I now know?

When I started this journey six years ago, it was with a sense that things were rapidly spiralling out of control and that I needed to drop everything and find out what if anything could be done to change this trajectory. If I was going to get behind something I needed to be sure we had a reliable compass. My education has come a long way since then, allowing me to connect the dots between the way we grow and process our food, the health of the planet and the health of humans. I am now very encouraged to find that there is a path forward by working with nature. My mind has been illuminated and the possibilities sparkle with light. I am no longer limited by my old beliefs. I think that is what the writer Waldo Emerson is saying in the quotation:

> "To the illuminated mind, the whole world sparkles with light."

I wake up each day aware that we are running out of time. I am conscious that we need to bring about a century of change within a decade. To tackle climate change, we need to mobilize all of society, like they did during the world wars. Somehow, the central focus of the population, the business of earning a living and raising children, must be dovetailed with actions needed to tackle the existential crises humanity faces. We need to develop narratives for how this can be accomplished.

We are not likely to achieve the big changes required without maintaining our social fabric, the glue that holds society together. This is currently under threat from a litany of social issues, including extreme economic inequality, systemic racism, high student loan debt, poor nutrition, chronic disease epidemics, the COVID-19 pandemic, climate refugees, and alternative truths. One possible solution is a Green New Deal[225][226][227] that is being discussed in many countries. Basically, we need a new pact between capital, work, and the planet. Economic progress must be conditioned on planetary limits. The needed shift to small biointensive regenerative farms offers the prospects of much greater employment for farmers producing highly nutritious food and connecting food production to human and environmental health needs.

Retirement allows Jackie and me, together with a small group of like minded individuals, to focus on educating ourselves and others and to tease apart the changes that need to occur in our society to shift to the new paradigm of regeneration. On the island where we live in British Columbia, we formed the Bowen Island 'FoodResilience' Society (BIFS). We are dedicated to exploring the principles of regenerative agriculture and trying them out in our own gardens and a large community garden. Our focus is to participate in a community effort to grow more food locally in ways that work with nature to regenerate the land resource and share the lessons learned with our families, our community, and the larger world. We are far from unique in this undertaking which is very encouraging because our institutions will only change to a new paradigm once a groundswell of public awareness emerges.

Alarms are going off around the world and more organizations, communities, and governments are declaring a climate emergency, but the awareness level is still not high enough to bring about the paradigm shift required.

Above all we need to:
- get off of fossil fuels
- acknowledge that the current agricultural war against nature is a war against ourselves
- focus on collaborating with nature's work force to grow our food in ecologically sound ways
- regenerate our soils using nature's natural cycles and regenerative principles
- increase awareness of the many ways to rapidly cool the planet by rebuilding the soil carbon sponge
- view nutritious food as medicine and pay attention to societal math that currently promotes sickness
- acknowledge hubris, a well known flaw in humans, and restore the precautionary principle
- move to holistic management to deal with environmental, social, and economic complexity
- increase awareness of the interconnections between soils, plants, microbes, animals, and the planet

soil health = plant health = planet health = human health

Once we turn that corner, I will again be able to accept a dinner invitation.

References

[1] Chris Arsenault, 'Only 60 Years of Farming Left If Soil Degradation Continues', Scientific American, 5 Dec. 2014.
https://www.scientificamerican.com/article/only-60-years-of-farming-left-if-soil-degradation-continues/

[2] Bibi van der Zee , 'UK is 30-40 years away from eradication of soil fertility, warns Gove', Guardian 24 Oct 2017.
https://www.theguardian.com/environment/2017/oct/24/uk-30-40-years-away-eradication-soil-fertility-warns-michael-gove

[3] Ingham, E.R., Moldenke, A. R., & Edwards, C. A., 'Soil Biology Primer', USDA Natural Resources Conservation Service.
https://www.nrcs.usda.gov/wps/portal/nrcs/main/soils/health/biology/

[4] Ingham, R. E.,Trofymow, J. A., Ingham, E. R., & Coleman, D. A.,'Interactions of Bacteria, Fungi, and their Nematode Grazers: Effects on Nutrient Cycling and Plant Growth', *Ecological Monographs,* Vol. 55, No. 1, Mar., pp. 119-140, 1985.

[5] Dr. Elaine Ingham's Soil Foodweb, 'Restoring Nature to the World's Soils'. https://www.soilfoodweb.com/

[6] 'The Green Revolution & Dr Norman Borlaug: Towards the "Evergreen Revolution"', AgBioWorld, 2011.
http://www.agbioworld.org/biotech-info/topics/borlaug/green-revolution.html

[7] Wikepedia, 'Norman Borlaug'. https://en.wikipedia.org/wiki/Norman_Borlaug

[8] Reay Tannahill, 'Food in History', New York Three Rivers Press, p. 336, 1988.

[9] 'Norman Borlaug', McGill School of Computer Science, 2007 Schools Wikipedia Selection.
https://www.cs.mcgill.ca/~rwest/wikispeedia/wpcd/wp/n/Norman_Borlaug.htm

[10] Dent, D and Cocking, E., 'Establishing symbiotic nitrogen fixation in cereals and other non-legume crops:: The Greener Nitrogen Revolution', Agric & Food Secur 2017, 6:7, DOI 10.1186/s40066-016-0084-2
https://agricultureandfoodsecurity.biomedcentral.com/track/pdf/10.1186/s40066-016-0084-2

[11] Darrin Qualman, 'Civilization Critical: Energy, Food, Nature, and the Future', Fernwood Publishing, April 2019.
https://fernwoodpublishing.ca/book/civilization-critical

[12] Darrin Qualman Blog, http://www.darrinqualman.com/canadian-net-farm-income/

[13] Norman Borlag (1970 Nobel Prize Laureate for Peace),
'Ending World Hunger. The Promise of Biotechnology and the Threat of Antiscience Zealotry', Plant Physiol. Vol. 124, 2000.
http://www.plantphysiol.org/content/124/2/487

[14] Jenifer Fraser, 'The World's Largest Mining Operation is Run by Fungi', Scientific American, 5 Nov. 2015.
https://blogs.scientificamerican.com/artful-amoeba/the-world-s-largest-mining-operation-is-run-by-fungi

[15] Landeweert, R. et al., 'Linking plants to rocks: ectomycorrhizal fungi mobilize nutrients from minerals',
Trends in Ecology & Evolution 16, no. 5, pp. 248-254, 2001.
https://www.researchgate.net/publication223028235_Linking_plants_to_rocks_Ectomycorrhizal_fungi_mobilize_nutrients_from_minerals

[16] Simard, S. W. et al., 'Net transfer of carbon between ectomycorrhizal tree species in the field', NATURE, VOL 388, p. 579, 1997

[17] Suzanne W. Simard, 'Nature's internet: how trees talk to each other in a healthy forest', TEDxSeattle, 2 Feb. 2017
https://www.youtube.com/watch?v=breDQqrkikM

[18] David R. Montgomery, 'Growing a Revolution: Bringing Our Soils Back To Life', professor of geomorphology
  a) paperback 10 July 2018, W. W. Norton & Company, New York – London
  b) YouTube video 17 May 2017. https://www.youtube.com/watch?v=c4p-kQ6D8aA

[19] Kurt Lawton, 'Economics of soil loss', Corn + Soybean Digest, 13 Mar. 2017.
http://www.cornandsoybeandigest.com/soil-health/economics-soil-loss

References

[20] Jason Bradford, 'One Acre Feeds A Person', Farmland LP, 13 Jan. 2012. http://www.farmlandlp.com/2012/01/one-acre-feeds-a-person/

[21] Melissa C. Lott, '10 Calories in, 1 Calorie Out - The Energy We Spend on Food' Scientific American, 11 August 2011.
https://blogs.scientificamerican.com/plugged-in/10-calories-in-1-calorie-out-the-energy-we-spend-on-food/

[22] Joe Cornelius, PhD, Program Director Advanced Research Projects Agency-Energy, 'Counting Calories', 28 Feb. 2017.
https://arpa-e.energy.gov/sites/default/files/2017_Cornelius_FastPitch_Final.pdf

[23] Woods, J. et al., Philos Trans R Soc Lond B Biol Sci., 365(1554), pp. 2991–3006, 2010, doi:10.1098/rstb.2010.0172

[24] Anne Helmenstine, 'Elements in the Human Body and What They Do', Science Notes, 20 May 2015 (updated on 7 February 2019)
https://sciencenotes.org/elements-in-the-human-body-and-what-they-do/

[25] F. Nielsen, 'Ultratrace Elements', Encyclopedia of Human Nutrition (Third Edition), pp. 299-310, 2013.
https://doi.org/10.1016/B978-0-12-375083-9.00270-1

[26] Thomas, David, 'A study on the mineral depletion of the foods available to us as a nation over the period 1940 to 1991', Nutrition and Health, 17(2), pp. 85-115, 2003, doi: 10.1177/026010600301700201. https://www.ncbi.nlm.nih.gov/pubmed/14653505

[27] Don M. Huber, keynote talk at the 35th National Pesticide Forum, April 2017.
https://www.youtube.com/watch?v=dwlTZRwlJYU&index=5&list=PLHS5IfcgFy5cXcht1lxwxIbAPekpNPmYQ

[28] André Picard, 'Today's fruits, vegetables lack yesterday's nutrition', The Globe and Mail, 6 July 2002.
https://www.theglobeandmail.com/life/todays-fruits-vegetables-lack-yesterdays-nutrition/article4137315/

[29] Roddy Scheer and Doug Moss, 'Dirt Poor: Have Fruits and Vegetables Become Less Nutritious?', Scientific American, 27 April 2011.
https://www.scientificamerican.com/article/soil-depletion-and-nutrition-loss/

[30] Reicosky, D. C. & Archer, D. W., Soil and Tillage Research, Vol. 94, Issue 1, pp. 109–121, 2007.

[31] David Johnson, New Mexico State University (NMSU), 'Rapid Carbon Sequestration'. https://www.youtube.com/watch?v=Fdh_j_KOmrY

[32] David Johnson, 'Managing Soils for Soil Carbon Sequestration: Dr David Johnson on Engineering Microbiology', to the Organic Consumers Association, 21 July 2016. https://www.youtube.com/watch?v=18FVVYKU9gs

[33] David Johnson, Living Soils Symposium Presentation 2019, Regeneration Canada
https://www.youtube.com/watch?v=aGiJt6e_gqQ&t=1400s

[34] Gabe Brown, 'Dirt to Soil', Chelsea Green Publishing, 2018.

[35] Charles Massy, 'Call of the Reed Warbler: A New Agriculture – A New Earth', Chelsea Green Publishing, 2018.

[36] Deborah R. Huso, 'The Many Faces of Success', Progressive Farmer, 16 Mar. 2015.
https://www.dtnpf.com/agriculture/web/ag/news/farm-life/article/2015/03/16/virginia-farmer-turns-300-a-month-2

[37] Gabe Brown, 'Treating the Farm as an Ecosystem with Gabe Brown Part 3', 25 Mar. 2017. https://www.youtube.com/watch?v=uUmIdq0D6-A

[38] Stanley, P. L. & Rowntree, J. E. et al., Agricultural Systems 162, p.249, DOI: 10.1016/j.agsy.2018.02.003, 2018.

[39] Bobby Gill, 'It's Not the Cow it's the How', TEDx talk 13 Feb 2020. https://www.youtube.com/watch?v=jKXgVK0TQ1A&t=921s

[40] Mariko Thorbecke & Jon Dettling, 'Carbon Footprint Evaluation of Regenerative Grazing At White Oak Pastures', Quantis 25 Feb. 2019.
https://blog.whiteoakpastures.com/hubfs/WOP-LCA-Quantis-2019.pdf

[41] Jeff Moyer, Andrew Smith, Yichao Ru, Jennifer Hayden, 'Regenerative Agriculture and the Soil Carbon Solution', Rodale Institute White Paper, Sept 2010. https://rodaleinstitute.org/wp-content/uploads/Rodale-Soil-Carbon-White-Paper_v9.pdf

[42] Teague, W. R. et al., 'The role of ruminants in reducing agriculture's carbon footprint in North America', Journal of Soil and Water Conservation, 71 (2), pp. 156-164, March 2016. https://doi.org/10.2489/jswc.71.2.156

[43] Machmuller, M. B. et al., 'Emerging land use practices rapidly increase soil organic matter', Nature Communications, 30 Apr 2015. doi: 10.1038/ncomms7995

References

[44] Andre Leu, 'Reversing Climate Change through Regenerative Agriculture ', Regeneration International 9 Oct. 2018.
https://regenerationinternational.org/2018/10/09/reversing-climate-change-through-regenerative-agriculture/

[45] Didi Pershouse, 'Understanding Soil Health and Watershed Function:A Teacher's Manual', latest update 2019,
https://www.didipershouse.com/understanding-soil-health-and-watershed-function.html

[46] Judith D. Schwartz, 'Water in Plain Sight, Hope for a Thirsty World', St. Martins's Press, New York, 2016.

[47] Dr. Christine Jones, 'Save our Soils — Dr. Christine Jones Explains the Life-Giving Link Between',
Carbon and Healthy Topsoil', interviewed by Tracy Frisch for ACRESUSA, 1 March 2015.
http://www.ecofarmingdaily.com/interview-sos-save-soils-dr-christine-jones-explains-life-giving-link-carbon-healthy-topsoil/

[48] Julienne Isaac, 'Farmers ahead of carbon curve', Grainviews, Jan. 29, 2016.
http://www.grainews.ca/2016/01/29/farmers-ahead-of-carbon-curve/

[49] Statistics Canada, 'Tillage practices used to prepare land for seeding', Table: 32-10-0408-01 (updated every 5 years), last update 1 Feb. 2021.
https://www150.statcan.gc.ca/t1/tbl1/en/tv.action?pid=3210040801&pickMembers
%5B0%5D=1.1&cubeTimeFrame.startYear=2011&cubeTimeFrame.endYear=2016&referencePeriods=20110101%2C20160101

[50] David R. Huggins and John P. Reganold, 'No-till: The Quite Revolution', Scientific American, pp. 70-77, July 2008.
https://www.ars.usda.gov/ARSUserFiles/20902500/DavidHuggins/NoTill.pdf

[51] Saskatchewan Soil Conservation Association Inc. (info@ssca.ca), ' SOIL CARBON POSITION PAPER (Version 1)',
Carbon Advisory Committee –August2017. http://ssca.ca/images/pdf/Soil-Carbon-Position-Paper-Version-1---August-2017.pdf

[52] Hershey, D. R., 'Digging Deeper into Helmont's Famous Willow Tree Experiment',
The American Biology Teacher, Vol. 53, No. 8, p. 458, 1991

[53] Feynman's explanation on YouTube. From the BBC TV series 'Fun to Imagine',1983, https://www.youtube.com/watch?v=ITpDrdtGAmo

[54] Latshaw, W.L. & Miller, E. C., 'Elemental Composition of the Corn Plant', Journal of Agricultural Research, Washington D. C.,
Vol. XXVII,11, pp. 845-861, 15 Mar. 1924. https://naldc.nal.usda.gov/download/IND43966853/PDF

[55] Latshaw, W. L., 'Elemental Composition of the Corn Plant', Journal of Agricultural Research, Washington D. C.,
Vol. XXVII, No. 11, Mar. 15, 1924.

[56] Mike McGinnis, 'U.S. Corn Yield Is Getting Bigger, USDA Says', Successful Farming, 10 Oct. 2019.
https://www.agriculture.com/news/crops/us-farmers-tol-grow-smaller-corn-soybean-crops-usda-says

[57] Donald Sparkes, 'Environmental Soil Chemistry', 2nd edition, Academic Press, 2002.

[58] Dr. Russell Briggs, 'Introduction to Soils for 345/545 Fall 2019', SUNY College of Environmental Sciences,
https://www.esf.edu/for/briggs/FOR345/erosion.htm

[59] Napieralski, S. A. el al., 'Microbial chemolithotrophy mediates oxidative weathering of granitic bedrock',
Proceedings of the National Academy of Sciences, 2019. www.pnas.org/cgi/doi/10.1073/pnas.1909970117

[60] Dr. Allen Williams, 'Restore Soil and Ecosystem Health with Adaptive Grazing', Quivira Conference, Albuquerque, 15-17 Nov. 2017.
https://www.youtube.com/watch?v=BwH6od6Jaq8   (at time 34.16)

[61] Science Learning Hub, 'Charles Darwin and earthworms', https://www.sciencelearn.org.nz/resources/22-charles-darwin-and-earthworms

[62] Gabe Brown, 'Regenerating our Lands: A Producers Perspective' TEDx GrandForks, 2016. https://www.youtube.com/watch?v=QfTZ0rnowcc

[63] Dr. Christine Jones, 'Digging deeper' soil biology forum, talk published 17 Dec. 2017. https://www.youtube.com/watch?v=EKHchVlwNRg

[64] Christine Jones, 'Soil Restoration: 5 Core Principles', Original article in Oct. 2017 issue of Acres U.S.A. magazine,
https://www.ecofarmingdaily.com/build-soil/soil-restoration-5-core-principles/

[65] Mulvaney, R. L., Khan, S. A., and Ellsworth, T. R., 'Synthetic Nitrogen Fertilizers Deplete Soil Nitrogen: A Global Dilemma for Sustainable
Cereal Production', J. Environ. Qual., 38, pp.2295–2314, 2009. doi:10.2134/jeq2008.0527

References

[66] Tom Philpott, 'New research: synthetic nitrogen destroys soil carbon, undermines soil health', grist 24 Feb. 2010.
https://grist.org/article/2010-02-23-new-research-synthetic-nitrogen-destroys-soil-carbon-undermines/

[67] Mulvaney, R. L., Khan, S. A., and Ellsworth, T. R., 'The Browning of the Green Revolution', GaiaCollege.ca Science News, 9 Mar. 2010.
https://www.gaiacollege.ca/moodle/mod/data/view.php?d=9&mode=single&page=502

[68] Ray S. K. et al., 'Recent patterns of crop yield growth and stagnation', Nature Communications, 3:1293, 2012. DOI: 10.1038/ncomms2296
https://www.nature.com/articles/ncomms2296.pdf

[69] White, J.W., 'Soil organic matter and manurial treatments'. J. Am. Soc.,Agron., 19, pp. 389–396, May 1927.
https://doi.org/10.2134/agronj1927.00021962001900050004x

[70] Albrecht, W.A., 'Variable levels of biological activity in Sanborn Field after fifty years of treatment. Soil Sci. Soc. Am. Proc. 3, pp.77–82, 1938. https://doi.org/10.2136/sssaj1939.036159950003000C0014x

[71] Kunkun fan et al., 'Suppressed N fixation and diazotrophs after four decades of fertilization', Microbiome, 7:143, 2019.
https://doi.org/10.1186/s40168-019-0757-8

[72] Ray, Deepak K. et al.,'Recent patterns of crop yield growth and stagnation', Nature Communications, 2012. DOI: 10.1038/ncomms2296
https://www.nature.com/articles/ncomms2296.pdf

[73] Wiesmeier, Martin, et al.'Stagnating crop yields: An overlooked risk for the carbon balance of agricultural soils?',
Science of The Total Environment, Vol. 536, pp. 1045-1051,1 Dec. 2015. https://doi.org/10.1016/j.scitotenv.2015.07.064

[74] Schauberger, Bernhard, et al., 'Yield trends, variability and stagnation analysis of major crops in France over more than a century',
www.nature.com/scientificreports/ (2018) 8:16865 | DOI:10.1038/s41598-018-35351-1

[75] Hochman, Ziv et al., 'Climate trends account for stalled wheat yields in Australia since 1990', Global Change Biology 23, 2071–208, 2017.
https://doi.org/10.1111/gcb.13604

[76] Allan Savory at TED 2013, 'How to fight desertification and reverse climate change'
http://www.ted.com/talks/allan_savory_how_to_green_the_world_s_deserts_and_reverse_climate_change

[77] Allan Savory with Jody Butterfield, 'Holistic Management: A Common Sense Revolution To Restore Our Environment',
published by Island Press, 3rd Edition, 2016.

[78] Jimi Eisenstein, 'What is Regenerative Agriculture', YouTube video, 20 Mar. 2020. https://www.youtube.com/watch?v=fSEtiixgRJl

[79] Phil Gregory, 'The Magic of Soil', YouTube, June 2017. https://www.youtube.com/watch?v=AWIL1YSf5ts

[80] Charles Griffith, Hugh Aljoe, & Russel Stevens, 'The Noble Foundation: D. Joyce Coffey Resource Management and Demonstration Ranch - Ten Years of Management and It Impact', 1996.
https://books.google.co.uk/books/about/The_Noble_Foundation.html?id=p0DkGwAACAAJ&redir_esc=y
Relevant figure found in presentation by Dr. Richard Teague, Green Cover Seed's 4th Annual Southern Soil Health Conference,
Nov. 28-29, 2017. https://www.youtube.com/watch?v=JX-bhRkVRzQ

[81] Alberta Lamb Proucers Fact Sheet, 'Precision Flock Management: Forage Growth and Intensive Grazing Basics'
https://www.ablamb.ca/images/documents/factsheets/PFM-Grazing-1-growth.pdf

[82] Allen, Myles R. et al.,
'A solution to the misrepresentations of CO2-equivalent emissions of short-lived climate pollutants under ambitious mitigation',
npj Climate and Atmospheric Science (2018) 16; doi:10.1038/s41612-018-0026-8 https://www.nature.com/articles/s41612-018-0026-8.pdf

[83] Robin Grieve, 'The Methane Mistake: NZ Carbon Emissions', June 2018.
https://www.youtube.com/watch?v=BOJdz_LgDBE&feature=youtu.be

[84] Allen, Myles R. et al., 'Climate metrics for ruminant livestock', Oxford Marton Program on Climate Pollutants, July 2018.
https://www.oxfordmartin.ox.ac.uk/downloads/reports/Climate-metrics-for-ruminant-livestock.pdf

References

[85] Ffinlo Costain, 'Ruminant agriculture can help us deliver net zero emissions', Posted on the British Veterinary Blog on October 15, 2019.
https://www.bva.co.uk/news-campaigns-and-policy/bva-community/bva-blog/ruminant-agriculture-can-help-us-deliver-net-zero-emissions/

[86] Mark Hume, 'Herbicide's collateral damage in B.C. forests under attack', The Globe and Mail, 24 July 2011, updated 3 May 2018.
https://www.theglobeandmail.com/news/british-columbia/herbicides-collateral-damage-in-bc-forests-under-attack/article625986/

[87] Jill English, 'Grooming forests could be making fires worse, researchers warn', CBC News, 24 Nov., 2019.
https://www.cbc.ca/news/canada/british-columbia/forest-fires-glyphosate-1.5366185

[88] Alexander, Martin E. 'Surface fire spread potential in trembling aspen during summer in the Boreal Forest Region of Canada', The Forestry Chronicle, Vol. 86, No.2, Mar/April 2010. https://pubs.cif-ifc.org/doi/pdf/10.5558/tfc86200-2

[89] 'Forest herbicide contributing to wildfires', CBC News, 21 Nov. 2019.
https://www.cbc.ca/news/thenational/forest-herbicide-contributing-to-wildfires-1.5369182

[90] Phil Gregory, 'Co-creating with nature: an exploration of Holistic Management'
https://www.phas.ubc.ca/~gregory/papers/HolisticManagement%20BowenGregory10Jun2018Sum.pdf

[91] The Global Savory Network, https://www.savory.global/our-network/

[92] Jonathan Watts, 'Third of Earth's soil is acutely degraded due to agriculture', Guardian, 12 Sep 2017.
https://www.theguardian.com/environment/2017/sep/12/third-of-earths-soil-acutely-degraded-due-to-agriculture-study#img-1

[93] Renee Cho, 'Can Soil Help Combat Climate Change?', State of the Planet, Earth Institute, Columbia University, |February 21, 2018.
https://blogs.ei.columbia.edu/2018/02/21/can-soil-help-combat-climate-change/

[94] Charles Massy, 'How regenerative farming can help heal the planet and human health', TEDxCanberra, 13 Nov 2018.
https://www.youtube.com/watch?v=Et8YKBivhaE

[95] Strassburg, Bernardo B. N., 'Global priority areas for ecosystem restoration', Nature, Vol. 586, 29 October 2020.
https://doi.org/10.1038/s41586-020-2784-9

[96] Moyer, Jeff, 'The Power of the Plate: The Case for Regenerative Organic Agriculture in Improving Human Health', Rodale Institute & The Plantrician Project; Sept. 2020.
https://rodaleinstitute.org/education/resources/power-of-the-plate-regenerative-organic-agriculture/

[97] 'Land to Market Program and Ecological Outcome Verification', Savory Institute, 15 Aug. 2018.
https://www.savory.global/wp-content/uploads/2018/08/0828_EOVDoc.pdf

[98] Tilman, D., Reich, P. B., & Isbell, F., 'Biodiversity impacts ecosystem productivity as much as resources, disturbance, or herbivory', Proceedings of the National Academy of Sciences; 2012; 109, 26, 10394-10397 2012.
http://cedarcreek.umn.edu/biblio/fulltext/Tilman-etal_PNAS_2012.pdf
YouTube video: https://www.youtube.com/watch?v=X0H_8awgpIA

[99] Lacis, A.A. et al., Atmospheric CO_2: Principal control knob governing Earth's temperature, Science, **330**, 356-359, 2010.
doi:10.1126/science.1190653

[100] Lacis, A.A., Hansen, J.E. et al., 'The role of long-lived greenhouse gases as principal LW control knob that governs the global surface temperature for past and future climate change', Tellus B: Chemical and Physical Meteorology, 65:1, 19734, 2013.
https://doi.org/10.3402/tellusb.v65i0.19734

[101] Walter Jehne, 'Climate Change: The Answer is Beneath Our Feet', Conversations from the Edge, a public talk from 3LM in Penrith, 2019.
https://www.youtube.com/watch?v=iD2DXBERTeg

[102] Walter Jehne, 'Restoring Water Cycles to Naturally Cool Climates and Reverse Global Warming',
Day one of an intensive workshop in 2016 at Lake Morey Resort in Vermont, USA. https://www.youtube.com/watch?v=K4ygsdHJjdI
Walter Jehne, 'Regenerating the Soil Carbon Sponge', Day two of an intensive workshop in 2016 at Lake Morey Resort in Vermont, USA.
https://www.youtube.com/watch?v=3nC6j80sLZo

References

[103] Walter Jehne and Didi Pershouse, 'The Soil Carbon Sponge, Climate Solutions and Healthy Water Cycles', 29 Apr 2018
   https://www.youtube.com/watch?v=123y7jDdbfY&t=528s

[104] Jehne, Walter, Proceedings of the Royal Society of Victoria; Melbourne, Vol. 126, (2014): 18-19.

[105] 'How Much More Will Earth Warm?', NASA Earth Boservatory, Jun 3, 2010;
   https://earthobservatory.nasa.gov/features/GlobalWarming/page5.php

[106] Franzluebbers, A. J., 'Stratification of Soil Porosity and Organic Matter', 2011
   In: Gliński J., Horabik J., Lipiec J. (eds) Encyclopedia of Agrophysics. Encyclopedia of Earth Sciences Series. Springer, Dordrecht

[107] Daisy Ouya, 'Cool insights for a hot world: trees and forests recycle water', Agroforestry, February 9, 2017
   http://blog.worldagroforestry.org/index.php/2017/02/09/cool-insights-hot-world-trees-forests-recycle-water/

[108] Read, P. L. et al., 'Global energy budgets and 'Trenberth diagrams' for the climates of terrestrial and gas giant planets',
   Quarterly Journal of the Royal Meteorological Society, 142: 703 – 720, 2016 B DOI:10.1002/qj.2704
   https://rmets.onlinelibrary.wiley.com/doi/pdf/10.1002/qj.2704

[109] David Battisti, 'Climate Change and Global Food Security', Atmospheric Science, University of Washington
   https://www.youtube.com/watch?v=YToMoNPwTFc

[110] Akshat Rathi, 'The world's first "negative emissions" plant has begun operation—turning carbon dioxide into stone', Quartz, 12 Oct. 2017
   https://qz.com/1100221/the-worlds-first-negative-emissions-plant-has-opened-in-iceland-turning-carbon-dioxide-into-stone/

[111] Kate Ravilious, 'The skies are alive with microbes that could be hijacking the weather', New Scientist, 16 Apr. 2016
   http://brent.xner.net/pdf/NewScient_clouds_April2016.pdf

[112] Wright, J. S. et al., 'Rainforest-initiated wet season onset over the southern Amazon', Proceedings of the National Academy of Sciences,
   Aug. 8, 2017 114 (32) 8481-8486   https://doi.org/10.1073/pnas.1621516114

[113] Ehn, M. et al., 'A large source of low-volatility secondary organic aerosol', NATURE, 02/2014, Vol. 506, Issue 7489

[114] 'How Trees Make Rain', Learning from Nature, January 15, 2020, This article has a nice list of references,
   https://www.learningfromnature.com.au/drought-proof-increasing-rainfall/#_ftn2

[115] Jimi Eisenstein, 'The Biotic Pump: How Forests Create Rain',
   https://www.youtube.com/watch?v=kKL40aBg-7E

[116] Jennifer Tsang, 'Snow Is Coming - What's That Have to Do with Microbes?' American Society for Microbiology, Jan. 11, 2019
   https://asm.org/Articles/2019/January/Snow-Is-Coming-Whats-That-Have-to-Do-with-Microbe

[117] Christner, Brent C. et al., 'Ubiquity of Biological IceNucleators in Snowfall',
   Science, 2008, Vol. 319, Issue 5867, DOI: 10.1126/science.1149757
   https://science.sciencemag.org/content/319/5867/1214

[118] Jim Morrison, 'Living Bacteria Are Riding Earth's Air Currents', Smithsonian.com, January 11, 2016
   https://www.smithsonianmag.com/science-nature/living-bacteria-are-riding-earths-air-currents-180957734/

[119] Walter Jehne, 'Regenerate the Earth', Healthy Soils Australia
   http://www.globalcoolingearth.org/wp-content/uploads/2017/09/Regenerate-Earth-Paper-Walter-Jehne.pdf

[120] Zhu, Xin-Guang et al., 'What is the maximum efficiency with which photosynthesis can convert solar energy into biomass?',
   Current Opinion in Biotechnology, Vol. 19, Issue 2, pp. 153-159, April 2008.

[121] Shelton Kroon, 'A Short History of Klipdrift', in 'Restoring South Africa's desertified Karoo',
   Managing Wholes, a project of the Soil Carbon Coalition;   https://managingwholes.com/klipdrift.htm

[122] 'SOIL ATLAS 2015' jointly published by the Heinrich Böll Foundation, Berlin, Germany, and the
   Institute for Advanced Sustainability Studies, Potsdam, Germany
   https://www.iass-potsdam.de/sites/default/files/files/soilatlas2015_web_english.pdf

[123] Bar-On, Y. M., Phillips, R., & Milo, R., 'The biomass distribution on Earth', PNAS 115 (25),pp. 6506-6511, June 19, 2018.
https://doi.org/10.1073/pnas.1711842115

[124] Rob Knight: 'How Our Microbes Make Us Who We Are', TED talk, 23 Feb. 2015. https://www.youtube.com/watch?v=i-icXZ2tMRM

[125] Bach, J. F., 'The Effect of Infections on Susceptibility of Autoimmune and allergic Diseases', N. England J. of Med., Vol. 347, 911, 2002.

[126] Marin J. Blaser MD, 'Missing Microbes: How the Overuse of Antibiotics is Fueling Our Modern Plagues', Henry Holt and Co., 2014.

[127] Mareen Ogle, 'Riots, Rage, and Resistance: A Brief History of How Antibiotics Arrived on the Farm', Scientific American, 3 Sept. 2013.
http://winewaterwatch.org/2018/01/glyphosate-5-billions-pounds-of-this-poison-sprayed-last-year/

[128] Don M. Huber,'The Failed Promises and Flawed Science of Genetic Engineering', Professor Huber interviewed by Dr..Mercola, 6 Oct.2013.
http://mercola.fileburst.com/PDF/ExpertInterviewTranscripts/Interview-DrHuber.pdf
https://www.youtube.com/watch?v=yx4UVhJcnpo

[129] W. Shaw & M. Pratt-Hyatt, 'The Importance of Testing for Glyphosate: The World's Most Widely Used Herbicide',
The Great Plains Laboratory, Inc., January 2017 issue of Townsend Letter
https://www.greatplainslaboratory.com/articles-1/2017/1/23/the-importance-of-testing-for-glyphosate-the-worlds-most-widely-used-herbicide

[130] Levine, H., 'Temporal trends in sperm count: a systematic review and meta-regression analysis', Human Reproduction Update, Vol.23, No.6
pp. 646–659, 2017. doi:10.1093/humupd/dmx022

[131] Kate Kelland, 'Sperm Count Dropping in Western World', Reuters on July 26, 2017.
https://www.scientificamerican.com/article/sperm-count-dropping-in-western-world/

[132] Bethell, C. D. et al., 'A national and state profile of leading health problems and health care quality for US children: key insurance disparities and across-state variations', Acad Pediatr May-Jun 2011;11(3 Suppl):S22-33. doi: 10.1016/j.acap.2010.08.011
https://pubmed.ncbi.nlm.nih.gov/21570014/

[133] Zach Bush, 'Eat Dirt! And Thrive',
2018 lecture connecting soil health, and the widespread use of pesticides in the US, to the rising rates of chronic disease
https://www.youtube.com/watch?v=HL6OPzQe9Is

[134] Zach Bush, 'Keynote: Hawaii Farmers Union United', 13 May 2020.
https://www.youtube.com/watch?v=G8W7twV8O54

[135] Nancy L. Swanson et al., 'Genetically engineered crops, glyphosate and the deterioration of health in the
United States of America', Journal of Organic Systems, Vol. 9, No. 2, 2014.

[136] Anthony Samsel, Stephanie Seneff, 'Glyphosate, pathways to modern diseases III: Manganese,
neurological diseases, and associated pathologies', Surgical Neurology International, Vol. 6
Issue: 1, p. 45, 2015. DOI: 10.4103/2152-7806.153876

[137] Mark Jeschke and Samantha Teten, 'Glyphosate-Resistant Weeds in North America', Dupont Pioneer, Vol. 10, No. 13, May 2018.
https://www.pioneer.com/CMRoot/Pioneer/US/Non_Searchable/agronomy/cropfocus_pdf/glyphosate-resistant-weeds.pdf
More source material: Heap, I. The International Survey of Herbicide Resistant Weeds. Online Internet. Wednesday, May 29, 2019.
Available www.weedscience.org

[138] Carey Gillam, 'It's Farmer v. Monsanto in Court Fight Over Dicamba Herbicide'., Sierra Club, 3 Feb 2020.
https://www.sierraclub.org/sierra/it-s-farmer-v-monsanto-court-fight-over-dicamba-herbicide

[139] Patricia Weiss and Gary McWilliams, 'U.S. Peach Grower Awarded $265 Million From Bayer, BASF in Weedkiller Lawsuit',
Reuters, 16 Feb. 2020.
https://www.reuters.com/article/us-bayer-dicamba-lawsuit/us-peach-grower-awarded-265-million-from-bayer-basf-in-weedkiller-lawsuit-idUSKBN20A0JJ

References

[140] Johnathan Hettinger ,'Federal Court Outlaws Controversial Herbicide Dicamba', Midwest Center for Investigative Reporting, 5 June 2020.
https://civileats.com/2020/06/05/federal-court-outlaws-controversial-herbicide-dicamba/

[141] United States Patent US 7,771,736 B2   https://patentimages.storage.googleapis.com/86/6d/8e/2d98b85f6574ef/US7771736.pdf

[142] Ken Roseboro,'Why Is Glyphosate Sprayed on Crops Right Before Harvest?', Eco Watch, Mar. 05, 2016.
https://www.ecowatch.com/why-is-glyphosate-sprayed-on-crops-right-before-harvest-1882187755.html

[143] Zach Bush in Episode 1 of 'GMOs Revealed' hosted by Dr. Patrick Gentempo, 26 Feb. 2019.

[144] Jeanne D'Brant, 'The Shikimate Pathway, The Microbiome, and Disease: Health Effects of GMOs on Humans'
https://d3n8a8pro7vhmx.cloudfront.net/yesmaam/pages/680/attachments/original/1466869052/GMO_Shikimate_pathway_gut_flora_and_health.pdf?1466869052

[145] Don M. Huber, 'Nutrition and Disease', Interview by Graeme Sait of nutri-tech – Part 2, 5 Dec. 2016.
https://blog.nutri-tech.com.au/don-huber-2/

[146] Jessica Stoller-Conrad, 'Microbes Help Produce Serotonin in Gut',Caltech, 9 April 2015.
https://www.caltech.edu/about/news/microbes-help-produce-serotonin-gut-46495

[147] Louise Hénault-Ethier, PhD et al, 'Alarming increase in the prevalence of autism: Should we worry about pesticides?' Sept. 5, 2019.
A Literature Review, Published by Autisme Montréal, Alliance pour l'interdiction des pesticides systémiques,
and the David Suzuki Foundation
https://davidsuzuki.org/wp-content/uploads/2019/09/alarming-increase-in-prevalence-of-autism-should-we-worry-about-pesticides.pdf

[148] Robert Lustig, 'Your Brain is Being Hacked Right Now', Reclaim Your Brain Conference: Keynote Presentation, May 8, 2019, London, UK
https://www.youtube.com/watch?v=XkjxKZ1dwB4

[149] Stice, E., Burger, KS; Yokum, S , 'Relative ability of fat and sugar tastes to activate reward, gustatory, and somatosensory regions',
Am. J. of Clinical Nutrition, 98(6):1377-84, 2013, DOI: 10.3945/ajcn.113.069443;   https://pubmed.ncbi.nlm.nih.gov/24132980/

[150] Robert Lustig, 'Sugar: The Bitter Truth', University of California Television, 30 July 2009. https://www.youtube.com/watch?v=dBnniua6-oM

[151] Alina Petre, 'Should You Peel Your Fruits and Vegetables?', 9 Dec. 2017. https://www.healthline.com/nutrition/peeling-fruits-veggies

[152] Kerns, C. E. et al., 'Sugar Industry and Coronary Heart Disease Research, a Historical Analysis of Industry Internal Documents',
JAMA Int. Med., 176(11):1680-1685, 2016. DOI: 10.1001/jamainternmed.2016.5394;
https://www.ncbi.nlm.nih.gov/pmc/articles/PMC5099084/

[153] Dehghen, Mashid & 35 other authors, 'Associations of fats and carbohydrate intake with cardiovascular disease and mortality in 18 countries from five continents (PURE): a prospective cohort study', The Lancet, Vol. 390, Issue 10107, P2050-2062, Nov. 04, 2017.
http://dx.doi.org/10.1016/S0140-6736(17)32252-3

[154] Robert Lustig interviewed by Dr. Joseph Mercola, 'Mind Hack — How Corporations Took Over Our Bodies and Brains', 10 Sept. 2017.
https://articles.mercola.com/sites/articles/archive/2017/09/10/processed-foods-health-effects.aspx
Transcript: https://mercola.fileburst.com/PDF/ExpertInterviewTranscripts/Interview-DrLustig-TheHackingOfTheAmericanMind.pdf

[155] Schwartz , Jean-Marc et al.,
'Effects of Dietary Fructose Restriction on Liver Fat, De Novo Lipogenesis, and Insulin Kinetic in Children with Obesity',
Gastroenterology, 153(3):743-752, 2017. doi: 10.1053/j.gastro.2017.05.043; https://www.ncbi.nlm.nih.gov/pmc/articles/PMC5813289/

[156] Aseem Malhotra, 'Covid 19 and the elephant in the room', European Scientist, 16 April 2020
https://www.europeanscientist.com/en/article-of-the-week/covid-19-and-the-elephant-in-the-room/

[157] Bornstein, S. R. et al., 'Endocrine and metabolic link to coronavirus infection', Nature Rev Endocrinol 16(6):297-298, 2020.
https://doi.org/10.1038/s41574-020-0353-9

[158] Robert Lustig, 'The Hacking of the American Mind', Kopriva Science Seminar Series, Montana State University, 8 Mar. 2018.
https://www.youtube.com/watch?v=tHYu8NlWDLU

References

[159] Zen Honeycutt & Henry Rowlands, 'Glyphosate Testing Full Report: Findings in American Mothers' Breast Milk, Urine and Water', Moms Across America & Sustainable Pulse, April 7, 2014. https://www.momsacrossamerica.com/glyphosate_testing_results

[160] CAS No. 1071-83-6 CAREX (CARcinogen Exposure), CAREX Canada is a multi-institution research project that combines academic expertise and government resources to generate evidence-based carcinogen surveillance program for Canada
https://www.carexcanada.ca/en/glyphosate/

[161] 'National Primary Drinking Water Regulations', United States Environmental Protection Agency,
https://www.epa.gov/ground-water-and-drinking-water/national-primary-drinking-water-regulations

[162] 'EU Drinking Water Legislation', European Glyphosate Environmental Information Source
http://www.egeis-toolbox.org/documents/4%20Drinking%20water%20legislation%20draft%20v3%20.pdf

[163] Lorraine Chow, Results of Glyphosate Pee Test Are in 'And It's Not Good News', Eco Watch, 12 May 2016.
https://www.ecowatch.com/results-of-glyphosate-pee-test-are-in-and-its-not-good-news-1891129531.html

[164] Eberbach, Philip, 'Applying Non-steady-state Compartmental Analysis to Investigate the Simultaneous Degradation of Soluble and Sorbed Glyphosate (N-(Phosphonomethyl)glycine) in Four Soils', Pestic. Sci., 52, pp. 229-240, 1998.
https://doi.org/10.1002/(SICI)1096-9063(199803)52:3<229::AID-PS684>3.0.CO;2-O

[165] Press release Wageningen University & Research, High levels of glyphosate in agricultural soil: 'Extension of approval not prudent.' 16 October 2017.
https://www.wur.nl/en/news-wur/Show/High-levels-of-glyphosate-in-agricultural-soil-Extension-of-approval-not-prudent.-.htm

[166] Silva, V. et al., 'Distribution of glyphosate and aminomethylphosphonic acid (AMPA) in agricultural topsoils of the European Union' Science of The Total Environment, Vol. 621, 1352-1359, 2018. https://doi.org/10.1016/j.scitotenv.2017.10.093

[167] John Kempf interviews Bob Kramer (USDA microbiologist, Prof. U. Missouri) 'Glyphosate in the environment', Regenerativ Agriculture Podcast, 19 Apr 2018.
https://www.youtube.com/watch?v=Y0eRsVeudQU

[168] Don Huber, Expert Witness Testimonial to the International Monsanto Tribunal an international civil society initiative, The Hague 2016.
https://de.monsantotribunal.org/upload/asset_cache/76342986.pdf?rnd=IT6q1h

[169] Daniel Arkin, 'Jury orders Monsanto to pay nearly $290M in Roundup trial', NBC News, 10 Aug. 2018.
https://www.nbcnews.com/news/us-news/jury-orders-monsanto-pay-290m-roundup-trial-n899811

[170] Sam Levin and Patrick Greenfield, 'Monsanto ordered to pay $289m as jury rules weedkiller caused man's cancer' Guardian, 11 Aug 2018. https://www.theguardian.com/business/2018/aug/10/monsanto-trial-cancer-dewayne-johnson-ruling

[171] Sam Levin, 'Monsanto found liable for California man's cancer and ordered to pay $80m in damages' Guardian, 27 Mar. 2019. https://www.theguardian.com/business/2019/mar/27/monsanto-trial-verdict-cancer-jury

[172] Sam Levin, 'Monsanto must pay couple $2bn in largest verdict yet over cancer claims', Guardian 13 May 2019.
https://www.theguardian.com/business/2019/may/13/monsanto-cancer-trial-bayer-roundup-couple

[173] John Fagan, PhD, Michael Antoniou, PhD, Claire Robinson, Mphil, 'GMO Myths and Truths',
2nd edition published in Great Britain in 2014 by Earth Open Source
http://livingnongmo.org/wp-content/uploads/2014/11/GMO-Myths-and-Truths-edition2.pdf

[174] Fagan, John et al., 'Organic diet intervention significantly reduces urinary glyphosate levels in U.S. children and adults', Environmental Research, Vol. 189, 109898, 2020. https://doi.org/10.1016/j.envres.2020.109898

[175] Hyland, Carl et al., 'Organic diet intervention significantly reduces urinary pesticide levels in U.S. children and adults', Environmental Research, Vol. 171, pp. 568-575, 2019. https://doi.org/10.1016/j.envres.2019.01.024

[176] Kendra Klein and Anna Lappé, 'Drop in pesticides in human urine after switching to organic food',15 Feb 2019.
https://www.theguardian.com/commentisfree/2019/feb/15/what-the-pesticides-in-our-urine-tell-us-about-organic-food

References

[177] Raymond A. Cloyd, 'What Are Pesticide Metabolites?', Greenhouse Product News, Dec. 2018.
https://gpnmag.com/wp-content/uploads/2018/12/GPN_1218_Dr.Bugs_.pdf

[178] Nicholas Kristof, 'Trump's Legacy: Damaged Brains', The New York Times, 28 Oct. 2017.
https://www.nytimes.com/interactive/2017/10/28/opinion/sunday/chlorpyrifos-dow-environmental-protection-agency.html

[179] Britt E. Erickson, 'No safe exposure to chlorpyrifos, EU regulators say', C&EN, 5 Aug. 2019.
https://cen.acs.org/environment/pesticides/safe-exposure-chlorpyrifos-EU-regulators/97/web/2019/08

[180] Health Canada Pesticide Product Information Database: https://pesticide-registry.canada.ca/en/index.html

[181] Danny Hakim, 'Glyphosate, Top-Selling Weed Killer, Wins E.U. Approval for 5 Years', NY Times 27 Nov. 2017.
https://www.nytimes.com/2017/11/27/business/eu-glyphosate-pesticide.html

[182] CBC News, 'Health Canada stands by approval of ingredient in Roundup weed killer', The Canadian Press 11 Jan. 2019.
https://www.cbc.ca/news/canada/saskatchewan/health-canada-herbicide-glyphosate-roundup-1.4975945

[183] 'Elements of Precaution: Recommendations for the Regulation of Food Biotechnology in Canada', An expert panel report on the future of food biotechnology prepared by The Royal Society of Canada at the request of Health Canada, Canadian Food Inspection Agency and Environment Canada, Conrad Brunk & Brian Ellis Co-Chairs, RSC Expert Panel on the Future of Food Biotechnology, Jan. 2001.
https://rsc-src.ca/sites/default/files/GMreportEN.pdf

[184] Peter Andre and Lucy Shatter, 'Genetically Modified Organisms and Precaution: Is the Canadian Government Implementing the Royal Society of Canada's Recommendations? Oct. 2004.
https://cban.ca/genetically-modified-organisms-and-precaution-is-the-canadian-government-implementing-the-royal-society-of-canadas-recommendations/

[185] Kosicki, Michael et al., 'Repair of double-strand breaks induced by CRISPR–Cas9 leads to large deletions and complex rearrangements', Nature Biotechnology, Vol. 36, pp. 765–771, July 2018. https://doi.org/10.1038/nbt.4192

[186] Carlson, Daniel F. et al. 'Production of hornless dairy cattle from genome-edited cell lines', Nature Biotechnology, Vol. 34, pp. 479–481, May 2016. https://doi.org/10.1038/nbt.3560

[187] Pimentel, David, 'Environment and Economic Costs of the Application of Pesticides, Primarily in the United States', Environment, Development and Sustainability, Vol. 7: pp. 229–252, 2005. DOI 10.1007/s10668-005-7314-2
https://link-springer-com.ezproxy.library.ubc.ca/content/pdf/10.1007/s10668-005-7314-2.pdf

[188] Lewis, W. J. et al., 'A total system approach to sustainable pest management',
Proc. Natl. Acad. Sci. USA, Vol. 94, pp. 12243–12248, Nov. 1997. doi: 10.1073/pnas.94.23.12243
https://www.pnas.org/content/94/23/12243

[189] Mullinix. K., 'Effect of cover crop on apple leafroller populations, leafroller parasitism and selected arthropods in an orchard managed without insecticides.', 2004, PhD Plant Science, University of British Columbia, Vancouver, British Columbia, Canada

[190] Mulinix, K., Isman, M. B., Brunner, J. F., 'Key and Secondary Arthropod Pest PopulationTrends in Apple Cultivated over Four Seasons with No Insecticides and a Legume Cover', Journal of Sustainable Agriculture, 34:6, pp. 584-594, 2010. DOI: 10.1080/10440046.2010.493363

[191] Rajinder Peshin & WenJun Jhang, 'Chapter 1 on Integrated Pest Management and Pesticide Use', in 'Integrated Pest Management', Pesticide Problems, Vol. 3, Editors David Pimentel & Rajinder Reshin, Springer 2014.

[192] Lauren Quinn, 'Illinois Regenerative Agriculture Initiative launches at University of Illinois', ACES News, 9 Oct. 2020.
https://aces.illinois.edu/news/illinois-regenerative-agriculture-initiative-launches-university-illinois

[193] Fresh Taste, https://freshtaste.org/

[194] Kirby Barth and Brian Frederick, '12 Orgs Highlighting the True Cost of Food', Sustainable Business, April 2019.
https://foodtank.com/news/2019/04/10-orgs-highlighting-the-true-cost-of-food

[195l] John Ikerd, 'Social and ethical values', in The True Cost Of American Food Conference, San Francisco, April 14-17, 2019, pp. 154 -156
http://sustainablefoodtrust.org/wp-content/uploads/2013/04/TCAF-report.pdf

References

[196] Fitzpatrick, Ian et al., 'The Hidden Costs of UK Food', revised edition 2019, Sustainable Food Trust,
https://sustainablefoodtrust.org/wp-content/uploads/2013/04/Website-Version-The-Hidden-Cost-of-UK-Food.pdf

[197] John Ikerd, 'Small Farms: Their Role in Our Farming Future', 1999 Keynote address to the MOSES Organic Farming Conference (transcript),
https://mosesorganic.org/wp-content/uploads/2020/03/John.Ikerd1999.keynote.pdf

[198] John Ikerd, 'Reclaiming the Future of Farming', 2020 Keynote address to the MOSES Organic Farming Conference (transcript),
https://mosesorganic.org/wp-content/uploads/2020/03/John.Ikerd_.2020-Keynote.pdf

[199] John Ikerd, 'The true cost of large-scale farming', in The True Cost Of American Food Conference, San Francisco, April 14-17, 2019, pp. 85-86; http://sustainablefoodtrust.org/wp-content/uploads/2013/04/TCAF-report.pdf

[200] Elizabeth and Paul Kaiser, Keynote Talk 2017 NOFA/Mass Winter Conference, 4 Jan. 2017,
https://www.youtube.com/watch?v=zAn5YxL1PbM

[201] Elizabeth and Paul Kaiser, Keynote Presentation 2019 Utah Farm Conference, 14 Jan. 2019.
https://www.youtube.com/watch?v=4lmV5Kx3Iw4

[202] Sonoma County Farm Bureau, 'Bee Friendly: Singing Frogs Farm Your Local No-Till CSA', 1 April 2015.
https://sonomafb.org/bee-friendly-singing-frogs-farm-your-local-no-till-csa/

[203] Jean-Martin Fortier, 'The Market Gardener: A Successful Grower's Handbook For Small-Scale Organic Farming',
New Society Publishers, 2014.

[204] Perrine & Charles Hervé-Gruyer, 'Miraculous Abundance: One Quarter Acre, Two French Farmers and Enough Food to Feed the World',
Chelsea Green Publishing, ISBN: 9781603586429, 2016.

[205] Felix de Tombeur et al., 'Effects of Permaculture Practices on Soil Physicochemical Properties and Organic Matter Distribution in Aggregates: A Case Study of the Bec-Hellouin Farm (France)', Frontiers in Environmental Science. October 2018.
https://doi.org/10.3389/fenvs.2018.00116

[206] Helfand, S. M. and Taylor, M. P. H., 'The Inverse Relationship between Farm Size and Productivity: Refocusing the Debate', Food Policy, Elsevier, In Press, 1 Oct. 2020. https://doi.org/10.1016/j.foodpol.2020.101977

[207] Charles Massy, 'Call of the Reed Warbler: A New Agriculture, A New Earth', Chelsea Green Publishing, 2018.

[208] The B.C. Food Security Task Force, 'The Future of B.C.'s Food System', Jan. 2020.
https://engage.gov.bc.ca/app/uploads/sites/121/2020/01/FSTF-Report-2020-The-Future-of-Food.pdf

[209] Jonathan Latham, PhD, and Allison Wilson, PhD, 'FDA Finds Unexpected Antibiotic Resistance Genes in 'Gene-Edited' Dehorned Cattle', Biotechnology, Health, News, August 12, 2019.
https://www.independentsciencenews.org/news/fda-finds-unexpected-antibiotic-resistance-genes-in-gene-edited-dehorned-cattle/

[210] 'COVID-19 and the crisis in food systems:Symptoms, causes, and potential solutions',
Communiqué by the International Panel of Experts on Sustainable Food Systems (IPES-Food), April 2020.
http://www.ipes-food.org/_img/upload/files/COVID-19_CommuniqueEN.pdf

[211] Allan Savory, 'Hope for Reversing Desertification and Climate Change', 21 Sept. 2018 at Stonewall Farm, Keene, New Hampshire,
https://www.youtube.com/watch?v=58Fu1_3EBVU

[212] 'Ecological Outcome Verified (EOV™)', The Science Inside Land to Market, Savory Institute
https://www.savory.global/land-to-market/eov/

[213] 'Preliminary Estimates of the Prevalence of Selected Underlying Health Conditions Among Patients with Coronavirus Disease 2019 — United States, February 12–March 28, 2020',
CDC COVID-19 Response Team, Morbidity and Mortality Weekly Report (MMWR) / April 3, 2020 / 69(13);382–386
https://www.cdc.gov/mmwr/volumes/69/wr/mm6913e2.htm?s_cid=mm6913e2_w

References

[214] Jeffrey Smith, 'Genetic Roulette: The Documented Health Risks of Genetically Engineered Foods',
Yes! Books, Chelsea Green Publishing, 2007.
> video Genetic Roulette: The Gamble of Our Lives (2012)
> video Secret Ingredients (2018)

[215] 'Pesticides in Canada', https://en.wikipedia.org/wiki/Pesticides_in_Canada

[216] Lelie T. Foster et al.,'Pesticide-free communities', BC Atlas of Wellness 2nd Edition, 2011, Department of Geography, University of Victoria
http://www.geog.uvic.ca/wellness/wellness2011/Ch5_38.pdf

[217] 'Pesticide Free BC', https://crecwebcom.files.wordpress.com/2016/11/pesticide-free-bc.pdf

[218] 'Montreal wants to ban use of herbicide glyphosate, calling it a public health issue', CBC News 5 Sept. 2019.
https://www.cbc.ca/news/canada/montreal/roundup-ban-montreal-glyphosate-1.5271133

[219] Damian Carrington, 'EU agrees total ban on bee-harming pesticides', Guardian 27 Apr. 2018.
https://www.theguardian.com/environment/2018/apr/27/eu-agrees-total-ban-on-bee-harming-pesticides

[220] Robert Lustig, 'Societal Math', 15 Oct. 2019. https://robertlustig.com/2019/10/societal-math/

[221] 'Non Communicable Diseases', World Health Organization, 1 June 2018.
www.who.int/news-room/fact-sheets/detail/noncommunicable-diseases

[222] 'Summary for Policymakers of IPCC Special Report on Global Warming of 1.5°C approved by governments',
Intergovernmental Panel on Climate Change (IPCC), Ocober 2018.
https://www.ipcc.ch/2018/10/08/summary-for-policymakers-of-ipcc-special-report-on-global-warming-of-1-5c-approved-by-governments/

[223] Ripple, W. J. et al., 'World Scientists' Warning of a Climate Emergency', BioScience, Vol. 70, Issue 1, pp. 8–12, January 2020.
https://doi.org/10.1093/biosci/biz088

[224] David Coady et al., 'Global Fossil Fuel Subsidies Remain Large: An Update Based on Country-Level Estimates',
International Monetary Fund working paper, May 2019. https://www.imf.org/external/index.htm enter WP/19/89 in search field

[225] Ronnie Cummins , 'Why the Food and Regeneration Movement Should Support a Green New Deal',
Regeneration International, 24 Feb. 2019.
https://regenerationinternational.org/2019/02/24/a-call-for-the-food-movement-to-get-behind-the-green-new-deal/

[226] Loujain Kurdi ,'The Pact for Green New Deal', Greenpeace Canada, 21 June, 2019.
https://www.greenpeace.org/canada/en/press-release/23529/the-pact-for-green-new-deal/

[227] Natalie Sauer, 'Spain's socialists win election with Green New Deal platform', Climate Home News , 29 April 2019.
https://www.climatechangenews.com/2019/04/29/spains-socialists-win-election-green-new-deal-platform/

[228] Elizabeth and Paul Kaiser , 'Singing Frogs Farm', 11 April 2020. https://www.localharvest.org/singing-frogs-farm-M23845

[229] Caitlin Dewey, 'Why farmers only get 7.8 cents of every dollar Americans spend on food', Washington Post, 2 May 2018.

[230] '2017 SUMMARY REPORT On Antimicrobials Sold or Distributed for Use in Food-Producing Animals',
U.S. Food and Drug Administration, Center for Veterinary Medicine, Dec. 2018. https://www.fda.gov/media/119332/download

[231] David Wallinga, MD & Avinash Kar, 'New Data: Animal vs. Human Antibiotic Use Remains Lopsided',
Natural Resources Defense Council, 15 June 2020.
https://www.nrdc.org/experts/david-wallinga-md/most-human-antibiotics-still-going-us-meat-production

[232] '2019 SUMMARY REPORT On Antimicrobials Sold or Distributed for Use in Food-Producing Animals,
U.S. Food and Drug Administration, Center for Veterinary Medicine, Dec. 2020. https://www.fda.gov/media/144427/download

[233] 'Estimated Agricultural Annual Use for Glyphosate', U.S. Geological Survey, 2017
https://water.usgs.gov/nawqa/pnsp/usage/maps/show_map.php?year=2017&map=GLYPHOSATE&hilo=L

References

[234] Casale, John & Lydon, John, 'Apparent Effects of Glyphosate on Alkaloid Production in Coca Plants Grown in Colombia',
J. Forensic Sci, Vol. 52, No. 3, pp. 573-578, May 2007. doi:10.1111/j.1556-4029.2007.00418.x

[235] 'Monsanto Agrees to Modify Roundup Ads in New York State', Associated Press. 25 November 1996.
https://apnews.com/article/d196b9a5bb54637a7b281760b0f7a966

[236] 'Monsanto guilty in 'false ad' row', BBC News, 15 October. http://news.bbc.co.uk/2/hi/europe/8308903.stm

[237] Sarah Jamal, 'Part 2: Even more reasons why we can't be confident in Health Canada's assessment that glyphosate is safe',
Communications Manager, Environmental Defence Canada, May 02, 2019.
https://environmentaldefence.ca/2019/05/02/part-2-health-canada-glyphosate-safe/

[238] Anthony Samsel & Stephanie Seneff, 'Glyphosate, pathways to modern diseases II: Celiac sprue and gluten intolerance',
Interdiscip Toxicol. 6(4), pp. 159–184, Dec. 2013. doi: 10.2478/intox-2013-0026
https://www.ncbi.nlm.nih.gov/pmc/articles/PMC3945755/

[239] 'Safeguarding with Science: Glyphosate Testing in 2015-2016', Canadian Food Inspection Agency (CFIA), Date modified: 11-04-2017.
https://www.inspection.gc.ca/food-safety-for-industry/food-chemistry-and-microbiology/food-safety-testing-bulletin-and-reports/executive-summary/glyphosate-testing/eng/1491846907641/1491846907985

[240] Gildas Meneu, 'Du glyphosate dans nos aliments', Radio-Canada, 20 February 2019.
https://ici.radio-canada.ca/nouvelle/1153714/glyphosate-pesticide-alimentation?fbclid=IwAR1p2pobGQ6uezln-8wlQqbWQAuu6DvBzrU6LMnMIqIBJtU5CWUbRLgx7BI

[241] 'What's in your lunch?', Environmental Defence Canada and Équiterre, September 2018.
https://environmentaldefence.ca/report/whats-in-your-lunch/

[242] Fasano, A., 'Zonulin, regulation of tight junctions, and autoimmune diseases', Ann NY Acad Sci, Vol. 1258, pp.25-33, 2012.

[243] Fasano, A., 'Leaky gut and autoimmune diseases', Clin Rev Allergy Immunol,Vol.42, pp. 71-78, 2012.

[244] Gildea, J. J. et al., 'Protective Effects of Lignite Extract Supplement on Intestinal Barrier Function in Glyphosate-Mediated Tight Junction Injury', Journal of Clinical Nutrition & Dietetics , Vol. 3, No. 1, 2017, DOI: 10.4172/2472-1921.100035

[245] Gildea, J. J. et al., 'Protection against Gluten-mediated Tight Junction Injury with a Novel Lignite Extract Supplement',
J Nutr Food Sci, Vol. 6: 547, 2016, doi:10.4172/2155-9600.1000547

[246] Rob Ward MD, 'Redox Signaling – Healing Science', 7 Oct 2013. https://www.youtube.com/watch?v=SzAmKkfXEXM

[247] New York State Senate Bill S6502A Prohibits the use of glyphosate on state property, Signed By Governor
https://www.nysenate.gov/legislation/bills/2019/s6502/amendment/a

[248] 'Chronic Disease in America', National Center for Chronic Disease Prevention and Health Promotion, Last reviewed 12 Jan. 2021.
https://www.cdc.gov/chronicdisease/resources/infographic/chronic-diseases.htm

[249] Peter Calamai, 'GM Food Report: Ottawa Rapped, Expert Study Considered Major Setback for Biotech Industry',
Toronto Star, February 5, 2001, http://www.lobbywatch.org/archive2.asp?arcid=3560

[250] 'IARC Monograph on Glyphosate', Mar. 2015, World Health Organization International Agency for Research on Cancer (IARC)
https://www.iarc.who.int/featured-news/media-centre-iarc-news-glyphosate/

[251] The Guardian, 'Roundup weedkiller 'probably' causes cancer, says WHO study', 21 Mar. 2015.
https://www.theguardian.com/environment/2015/mar/21/roundup-cancer-who-glyphosate-

[252] David Thomas, 'The Mineral Depletion of Foods Available to US as A Nation (1940–2002) -a review of the 6th Edition of McCance and Widdowson', Nutrition and Health, Vol. 19, pp. 21-55, 2007, doi: 10.1177/026010600701900205.

[253] Davis, D. R. et al, 'Changes in USDA food composition data for 43 garden crops, 1950 to 1999', J Am Coll Nutr.,23(6), pp. 669-682, 2004.
doi: 10.1080/07315724.2004.10719409

References

[254] Andrew Purvis, 'It's supposed to be lean cuisine. So why is this chicken fatter than it looks?', The Guardian Observer, 15 May 2005. https://www.theguardian.com/lifeandstyle/2005/may/15/foodanddrink.shopping3

[255] Davis, D. R., 'Declining Fruit and Vegetable Nutrient Composition: What Is the Evidence?', Hortic. Science, Vol. 44, 1, pp.15-19, 2009, DOI: 10.21273/HORTSCI.44.1.15, https://journals.ashs.org/hortsci/view/journals/hortsci/44/1/article-p15.xml

[256] Davis, D. R., 'Nutritional Declienes in Foods: 8 Key Points', 2019 Lecture published on YouTube 22 June 2019. https://www.youtube.com/watch?v=mcM4FrXKDRc

[257] Robb Wolf, 'Guest Interview: Will Harris of White Oak Pastures', 20 Feb, 2020. https://www.youtube.com/watch?v=iN9fATbVGA4&t=1480s

[258] Eric Lipton, 'E.P.A. Chief, Rejecting Agency's Science, Chooses Not to Ban Insecticide', New York Times, March 29, 2017. https://www.nytimes.com/2017/03/29/us/politics/epa-insecticide-chlorpyrifos.html

[259] Sylvain Charlebois et al., 'Biotechnology in Food: Canadian Attitudes towards Genetic Engineering in both Plant-and Animal-based Foods', Dalhousie University, 2018. https://cdn.dal.ca/content/dam/dalhousie/pdf/management/News/News%20%26%20Events/Dalhousie%20GMO%20Food%20Study%202018%20(EN).pdf

[260] Porter, Stephanie S., Sachs, Joel L., 'Agriculture and the Disruption of Plant–Microbial Symbiosis', Trends in Ecology & Evolution, Vol. 35, No. 5, pp. 426-439, May 2020. DOI: 10.1016/j.tree.2020.01.006

[261] Holman, H. R.,'The Relation of the Chronic Disease Epidemic to the Health Care Crisis', ACR Open Rheumatology Vol. 2(3), pp.167–173, Mar. 2020. DOI: 10.1002/acr2.11114 https://www.ncbi.nlm.nih.gov/pmc/articles/PMC7077778/

[262] EPA Press Office, 27 Oct. 2020, EPA Announces 2020 Dicamba Registration Decision, https://www.epa.gov/newsreleases/epa-announces-2020-dicamba-registration-decision

[263] Debnath, Sovan et al., 'Are the modern-bred rice and wheat cultivars in India inefficient in zinc and iron sequestration?' Environmental and Experimental Botany 189, (2021) 104535, https://doi.org/10.1016/j.envexpbot.2021.104535

Acknowledgements

Since our retirement nearly 20 years ago, my wife Jackie has been hoping that we might combine our strengths on a meaningful endeavour to leave the world a better place for our children, grandchildren, and now great grandchildren. That day finally happened in 2015 when I turned my focus from astrophysics to an investigation of the way we grow our food and it's impact on environmental, human, and planet health. The previous year I had helped her realize her goal of a second garden based on a Hugelkultur design. In one year I went from focusing on an engineering challenge to minimize the garden construction cost to researching the latest soil biology practices that could regenerate the soil fertility and heal the planet. She is all about local meaningful action on food resilience in this era of climate change and I am all about research and understanding of a better global way forward. In the process of writing this book Jackie has been my biggest champion, always excited and willing to provide valuable feedback on this labour of love.

I have also received very helpful feedback from family members, relatives, friends and colleagues. In this regard I would like to acknowledge Jeff and Debbie Gregory, William Ostrander, Rabia Wilcox, and Helen and Patrick Sills for providing valuable comments on earlier drafts. Two of my physics colleagues Peter Martin and Jess Brewer provided very helpful editing and challenging questions for closer examination. I have enjoyed the encouragement of Allan Savory and Jody Butterfield on this journey and benefited greatly from interviews with Harold Steves and Dr. Kent Mullinix. Many other champions of the fields I have explored, including Dr. Don Reicosky, Gabe Brown, Professor Don Huber, Dr. Robert Lustig, Professor John Ikerd, Professor Bruce Lanphear, Darrin Qualman, and Elizabeth Kaiser have willingly responded to my questions or reviewed chapters pertaining to their own work. I also would like to acknowledge Dr. Elaine Ingham, a key pioneer in soil biology for the knowledge I acquired from her four foundational courses on the Soil Food Web.

Alphabetical Index

A
agricultural economics 53
Alberta Lamb Producers 33
Albrecht, W. A. 29
alkaloids 70
Alkaloids 70
Allan Savory 31
alzheimer's 71
amino acids 68, 69, 70
AMPA 69
antibiotics 64, 65, 66, 78
antimicrobial 65
archaea 21, 63
arthropod population dynamics 52
asbestos forests" 37
ascorbic acid 27
aspen 37
asthma 70
atmospheric physics 47
autism 71

B
B.C. Agricultural Land Reserve (ALR) 53
bacteria. 10, 11, 12, 15, 21, 23, 25, 29, 47, 63, 64, 66, 68, 69, 85
barter system 12, 14, 24
Berry, Wendell 90
bifidobacterium 78
biodiversity 12, 29, 40, 41, 42, 43, 89
biofertilizer 42, 49, 90
biology 44, 47, 85
biotic glues 10, 17
biotic pump 47
birch 37
bison 32
blood-brain barrier 67
blueberry bushes 37
Borlaug, Norman Ernest 6, 8
boron 25
brain 63, 69, 74, 75, 77
brain fog 3
bread 17
breeding plants 27
British Veterinary Association 36
broadfork 58
broccoli 27
Brown, Gabe 20, 23, 42
Bush, Zach 70, 78
Butterfield, Jody 37

C
calcium 26, 27
Caligari, Ademir 42
cancer 66, 70, 71, 75, 77, 78, 80, 81, 85
carbohydrate 14, 25, 27, 45, 75
carbon dioxide 15, 21, 29, 49
carbon footprint 33
carbon sequestration 15, 33, 40, 44, 47, 49
cattle 24, 32, 33, 35
causation 72
cereal crop 6
cereal crops 29
cheese 26
chelator 12, 14, 65
chemical-intensive agriculture 12, 88
chlorpyrifos 77
chronic disease. 64, 65, 67, 71, 72, 74, 76, 78, 85, 88, 89, 90
circular flows 7
climate change 5, 17, 30, 35, 44, 92
climate change 5
climate emergency 92
clothianidin 77
clouds 44, 46, 47
codling moth 52
Coleman, David C 13
Coleman, Eliot 60
collapse 5

collapse of public health...72
compost...11, 90
conifer..37
cooling..36, 44, 45, 46, 47, 85
copper..12, 25, 26, 27, 28
corn.......................................21, 22, 23, 29, 46, 74, 75
correlation..71
cortisol..76
COVID-19..75, 89
cows..35
Crawford, Michael...26
crop yield stagnation..29
cultivar...28
Cumberland, Rod..37
D
Daniels, Lori..37
Davis, Donald R..27
DDT..78
deep-organic...60
degree of correlation..71
dementia...71
Dennis, Neil...32
desertification.............................5, 31, 33, 34, 46, 90
Desroches, Maud-Hélène..59
diabetes...71, 74, 75
dicamba..65
dilution effect...27
dirt to soil..23, 24, 42
DNA..63
dopamine..72, 75, 76
Douglas fir...37
draw down...47
drinking water..77, 82
Druker, Steven M..82
E
Earth's biosphere...91
earthworm..12, 42
Ecological Outcome Verification...............................41
economic externalities...55
ecosystem...29, 38, 41, 43
ecosystem restoration...40

Einstein, Albert..1
elements..
    trace...12, 26
    ultratrace..12
elimination diet..3
Emerson, Waldo...92
enterococcus faecalis...78
EPA, U.S. Environmental Protection Agency....77, 78, 79
epidemic.......................................64, 67, 72, 73, 78, 88, 89, 90
European Food Safety Authority.................................77
evapotranspiration..15, 45
externalized cost..72
externalized costs...55, 57
extinction..1, 8, 40
extract, compost..11
exudates..12, 14, 24, 25, 42, 45
F
fallow..46
farm income crisis..7
Farmer's Footprint...90
Fasano, Alessio..67
fat..27
FDA...75
feedlot..33, 34
fertilizer..
    nitrogen..12, 22, 29
    phosphorus..12
    potassium...12
Feynman, Richard Phillips..21
fibre...74
flawed science...80
flour...17
folate...27
food as medicine...93
food insecure...56
food security..88
food sensitivities..3, 67
Fortier, Jean-Martin...59
fossil fuels...................................10, 35, 40, 88, 89
Four Season Farm..60
Fresh Taste...53

fructose 74, 75
fruit 3, 26, 74
fungi 10, 11, 12, 23, 24, 25, 63
G
gastrointestinal (GI) tract 67
General Mills 33
genetic dilution 27
genetic engineering (GE) 12, 64, 71, 72, 80, 82
genetically modified 82
genetically modified organism (GMO) 83
global thermostat 44
global warming 35, 36, 88
global warming potential 36
glucose 74, 75
glues 14, 24, 45
glycogen 74
glyphosate ... 37, 64, 65, 66, 69, 70, 71, 72, 74, 77, 78, 79, 80, 82, 89
GMO 60, 66, 83, 89
goats 33
grains 29, 34, 66
granulovirus 52
grazing 5, 25, 32, 33, 34, 38, 46
Green New Deal 92
green revolution 1, 6, 7, 11, 22, 28, 29, 50, 51, 85, 88
greenhouse blanket effect 44
greenhouse gas (GHG) ... 8, 15, 16, 27, 29, 35, 36, 40, 44, 45, 47, 55, 57, 90, 91
greenhouses 60
gross income 60
gross revenue 58
gut microbiome 64
H
hacking 75, 76
Haney, Rick 24
Harris, Will 33
health 42
Health Canada 77, 82
hepatitis C 72
herbicide 12, 14, 37, 64, 65, 80
herbivores 31, 33, 34, 35, 36, 49, 88, 90

Heuvé-Gruyer, Perrine & Charles 60
hidden cost 56
hierarchy 12, 63
high fructose corn syrup (HFCS) 74
holistic management 37, 38, 42, 90, 91
holistic planned grazing 33, 35
hormone 74
Huber, Don M. 69, 72, 78, 80
hubris 88, 93
human gut 71, 72
humus 17
hybridization 27
hypertension 71
I
Ikerd, John E. 53, 55, 56
immune system 69
Impossible Burger 33
Impossible Foods 33
industrial food system 51, 54, 56, 88
Industrial Revolution 40
infiltration 24, 40, 41, 42
inflammatory bowel disease 71
Ingham, Elaine R. 5, 11, 13, 14, 23
Ingham, Russell E. 13
insulin 74, 75
insulin resistance 74, 75
Integrated Pest Management (IPM) 51
inverse correlation 27, 71
iron 12, 27, 28
J
Jean Baptista van Helmont 21
Jehne, Walter 44, 48, 49
Jones, Christine 17, 25
K
Kaiser, Elizabet & Paul 58
kidney failure 71
kingdoms of life 63
KPU Institute for Sustainable Food Systems ... 53
Krüger, Monika 78
Kwantlen Institute for Sustainable Food Systems ... 53
Kwantlen Polytechnic University 50, 53

L

La Ferme du Bec Hellouin 60
Lacis, Andrew 44
lactobacillus 78
Land to Market 41
Lanphear, Bruce 78
latent heat 45
lead 78
leaf stomata 46
leaky gut 3, 66, 67
leptin 74
Les Jardins de la Grelinette 59
Lewis, W. J 51
life cycle assessment (LCA) 33
linear systems 7
liver 74, 75
low input sustainable agriculture (LISA) 50
Lustig, Robert H 74, 75, 76, 90

M

magnesium 12, 22, 23, 25, 26, 27
maize 27, 29, 46
malathion 77
Malhotra, Aseem 75
mammals 69
mandatory labelling 83
manganese 12, 25, 27
Massy, Charles 23, 24, 40
McCance, R. A 26
meat 26, 35, 36
melatonin 72
metabolite 77
methane 33, 35, 36
microbes 10, 11, 12, 14, 15, 17, 21, 22, 23, 24, 25, 28, 37, 42, 45, 46, 47, 49, 63, 64, 69, 72, 89
microbiome 64
microbiome, human 63, 64, 78, 88, 90
microscopic predators 10, 12
migraine 3, 4
milk 26
mineral content 26
mining 10, 12, 25
Miraculous Abundance 60
moisture content 23
monoculture 6, 8, 42
Montgomery, David R 10
Morrow Plots 29
Mullinix, Kent 50
multiple sclerosis 71
mycorrhiza 10, 23, 24, 25, 28

N

NASA 44, 47
native soils 10, 17
natural capital 37
naturopath 3
negative correlation 71
neonicotinoids 78
net revenue 58
neurotransmitters 63, 69, 72, 75, 76
niacin 27
Nichols, Kris 24
Nicolas-Théodore de Saussure 21
nitrogen fertilizer 29
nitrogen fixation 28, 29
nitrous oxide 29, 47
nucleating clouds 47
nutrient content 27
nutrient dilution 27, 28
nutrition 10, 21, 26, 27, 92
nutritional therapist 26

O

obesity 71, 75
oranges 12

P

P value, statistical significance 71
pancreas 74, 75
paradigm 1, 31, 38, 55, 89, 91, 92
parkinson's 71
Pershouse, Didi 17
pesticide 3, 12, 29, 40, 77, 78, 79, 80, 82, 90
phenylalanine 70, 72
pheromone 52
phosphorus 22, 23, 25, 26, 27, 69

photosynthesis......................12, 21, 25, 27, 35, 42, 45, 49
phytochemicals..28
pine...37
planetary limits..92
Plantrician Project...41, 90
plow...10, 15
Pollan, Michael..3
Polyface Farm..24
population stability..12
Porter, Stephanie...28
positive correlation..71
potassium...22, 23, 25, 26
potatoes...27
precautionary..83, 90
precautionary principle....................................88, 93
predators..32, 34
processed food...28, 74, 82, 90
protein...27, 67, 69
public health..72, 75, 82
pyrethroids..78
Q
Qualman, Darrin..6, 7, 51
Quantis...33
R
R value..71
radiative forcing..45
rain......................17, 21, 30, 42, 63, 64, 65, 69, 74, 75, 77
rainfall...17
raspberries...27
rebalancing..15, 44
reductionist management...............................38, 90, 91
regenerative agriculture 23, 35, 40, 42, 44, 48, 49, 58, 85, 88, 89, 90, 92
Regenerative agriculture...40
Reicosky, Don C...15, 19
renal disease..71
resilience...12
resistance...38, 65, 74, 75
rewilding...40
rhizobia...28
riboflavin...27

rice...3, 27, 28, 29, 30
Robbins, John..3
Rodale Institute..41, 90
rototiller..58
Roundup®..64, 66, 67, 69, 77, 80
Royal Society of Canada...................................82, 83
S
Sachs, joel...28
Salatin, Joel...23, 24
Savory Institute...37, 38, 41, 46
Savory, Allan.............................24, 35, 37, 38, 42, 89
serotonin..72, 76
sheep...33, 35, 58
shikimate pathway.......................................69, 70, 85
Shiva, Vandana..88
sigmoid growth curve..31, 33
silver bullet..37, 52, 88
Simard, Suzzane..10
Singing Frogs Farm...58
small farms..61
Smith, David J...47
Smith, Jeffrey M..89
social fabric...92
soil...
    biology..................5, 12, 13, 17, 22, 31, 40, 87, 88
    carbon sequestration.......................................40
    carbon sponge.....17, 22, 36, 41, 44, 45, 47, 85, 90, 93
    cathedrals..45
    degradation..5, 22, 29, 85
    degradation,..41
    erosion rate...10
    fertility...5, 15
    health...5, 24, 41, 49, 88
    internet..25
    microbiome..63, 88
    microscope..11
    minerals..21, 22, 25, 69
    organic carbon....................................15, 17, 19, 29, 45
    structure..11
    topsoil..17
soil food web..5, 28, 63

soil science, soil biology, and ecology 58
soil to dirt 10
specialization, simplification, standardization, and
mechanization 56
statistics for California vegetable farms 58
Steiner, Rudolf 24
Steves, Harold 53
Strassburg, Bernardo N. B. 40
stroke 71
sucrose 74
sugar 12, 14, 21, 25, 49, 74, 75
sunlight energy 12
Sunnybrae Acres 32
sustainable 55
Swanson, Nancy L. 71
symbiotic communities 28
symptoms 72, 73
synthetic fertilizers 24, 29, 42
systemic pesticide 77

T

tea, compost 11
technological treadmill 51
The Market Gardener 59
thiamin 27
Thomas, David 26
tight junctions 67
till 10
Tilman, David 42, 43
trait 27, 28, 83
transpire 45
Trofymow, J. A. 13
True Cost Accounting 55
tryptophan 70, 72
tyrosine 70, 72

U

UBC Faculty of Land and Food Systems 52
UK agriculture 36
UK Sustainable Food Trust 56
UN Food and Agriculture Organization 5
unintended consequences 37
University of British Columbia Faculty of Agriculture. 52
University of Illinois Regenerative Ag Initiative 53
USDA Agricultural Research Services 16, 24, 51
USDA Natural Resources Conservation Service 5

V

vegetable 26
vegetable farms 58
virus 63
vitamin A 27
vitamin E 27
vitamins 69
Vivre avec la terre 60
voids 17, 45

W

water 47
water cycle 36, 41, 44, 45, 47, 65
water soluble 65
water vapour 44
weed killer 80
wheat 27, 28, 29, 66
White Oak Pastures 33
White, J. W. 29
Widdowson, E. M. 26
wildebeest 32
wildfire 37
Williams, Allen 23
willow tree experiment 21
Wood Wide Web 10, 11

Z

zinc 12, 25, 26, 27, 28
Zonulin 67

www.ingramcontent.com/pod-product-compliance
Lightning Source LLC
Chambersburg PA
CBHW040543220526
45473CB00016B/3005